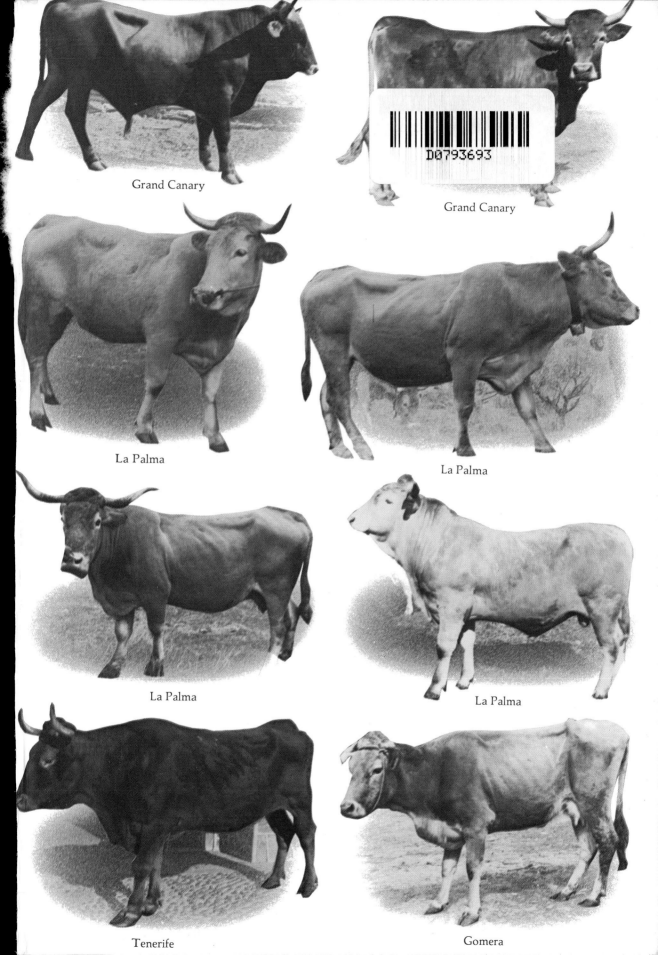

Grand Canary

Grand Canary

La Palma

La Palma

La Palma

La Palma

Tenerife

Gomera

THE CRIOLLO

Spanish Cattle in the Americas

THE CRIOLLO

Spanish Cattle in the Americas

John E. Rouse

University of Oklahoma Press : Norman

By John E. Rouse

World Cattle, I: Cattle of Europe, South America, Australia, and New Zealand
(Norman, 1970)
World Cattle, II: Cattle of Africa and Asia (Norman, 1970)
World Cattle, III: Cattle of North America (Norman, 1973)
The Criollo: Spanish Cattle in the Americas (Norman, 1977)

Library of Congress Cataloging in Publication Data

Rouse, John E
    The Criollo: Spanish cattle in the Americas.

    Bibliography: p.
    Includes index.
    1. Criollo cattle.   2. Cattle—America.   I.   Title.
SF199.C74R68   636.2'7   76–62506
ISBN 0–8061–1404–5

To Karen, my granddaughter

There's a Legion that never was listed,
That carries no colours or crest,
But, split in a thousand detachments,
Is breaking the road for the rest.

Kipling

PREFACE

Fourteen years ago I began my travels to see all types of cattle in the world in their native habitats. Although my objective was not completely realized, I did manage to see most of the different breeds in their respective homelands.

On many treks through Latin America I kept encountering an interesting cow which displayed a Jersey-tan colored hair coat. She was seen only in areas where true descendants of the old Spanish cattle still persisted. A distinctive beast, smaller in size in the mountainous areas of Central America than on a Caribbean island or in the Bolivian rain forest, she always had similar characteristics. Her barrel-shaped body, long, upswept horns, and short, fine hair coat denied kinship with the familiar cow of the Isle of Jersey. She was a Criollo. Throughout Latin America, "the cattle of the country," if they cannot be related to a modern breed, are known as Criollo—the designation given to man or beast considered "native" to the land.

In widely separated areas where the Criollo had been selectively bred for draft or milking ability, as in Colombia, Cuba, Venezuela, and Costa Rica, she became a larger animal, but retained many of the same characteristics as her sister who was the product of natural selection. The relationship was inescapable. I became intrigued by the persistent appearance of this Criollo cow in countries of Spanish heritage, and curiosity led me to seek her background.

My first discovery was that, while there was no question about the direct relationship between the Criollo cattle in different lands, this tie went back nearly five centuries. Not only my Criollo cow but the entire cattle population of Latin America, until the mid-nineteenth century, traced to a few hundred head of Spanish cattle brought over in Columbus' day from Andalusia and the Canary Islands. These included a sprinkling of black cattle, and probably a few black-and-white, which color patterns occasionally show up in the few survivors seen today.

The next revelation was the paucity of factual information on any kind of cattle from the Middle Ages down to fairly recent times. Meaningful records of how man husbanded his animals and the different types that existed were not transcribed until the landed gentry of England, at the close of the eighteenth century, began to develop their breeds of cattle and post their herd books.

In tracing the Criollo cow to her source, I followed a very cold and dim trail. History records the storms and calms encountered on voyages to the Indies, the sight of a bird, or the ill humor of the sailors, but mentions only incidentally that cattle were landed, ". . . along with plants, horses, asses, and pigs"—and fails to record how many. But from odd bits of evidence from many sources a reasonably authentic trail or chronology can be described of Spanish cattle from Columbian days until they became the Criollo in the Americas. The process reveals many highlights of history illuminating the interdependence of men and cattle in the New World.

The colony Columbus had planted in the Indies hung by a thread for the first decade until a breeding herd of cattle had been established. Food shortages that at times reached the point of famine led to great discontent and finally mutiny, until cattle were available for draft power and beef. As conquest proceeded from island to island and then on to the mainland, settlements and colonization were always secured by a base of cattle. As the conqueror's habitations became stabilized, the surplus from his herds spread out to virgin lands and founded wild populations which supported future advances.

When the age of conquest drew to a close at the end of the eighteenth century, it was followed by the wave of independence that engulfed all of Latin America in the following three decades. Again the Spanish cattle were to play a dominant role. Simón Bolívar, the greatest liberator of them all, rode to glory on the Criollo cow. Millions of these hardy beasts had spread over the Venezuelan llanos. Their hides and flesh bought the armament and fed the armies with which the general almost established a United States of South America.

The cattle dominions that flourished on the virgin grasslands from the Argentine to Canada were built on the fecundity of the Spanish cow. These vast ranges knew only the Criollo cattle until after the middle of the nineteenth century. In the temperate zones—Argentina, Uruguay, and western North America—bulls of the British breeds then transformed these Spanish cattle into herds of Shorthorn, Hereford, and later Angus within several cattle generations. Somewhat later the Criollos in the tropics were converted to Zebu-type cattle by the introduction of humped bulls from India.

True descendants of the old Spanish cattle have now all but disappeared. They are gone completely from the range lands of the hemisphere. The Criollo of the South American llanos and the mountains of Central America—true survivors of the fittest who have known little of man's care—can now be counted in the few thousands and cannot be expected to outlive another decade. A few tropical areas, in Cuba, Colombia, Venezuela, and Central America, still have small "islands" of improved Criollo cattle which have resulted from the breeding control which man has imposed. These beautiful and productive animals are little appreciated by the cattlemen and animal scientist of their homeland. Usually in the hands of an experiment station or university, these cattle and their survival are dependent on the varying winds of government.

The Criollo cow has been consistently ignored by historians, yet one segment of her progeny, the Longhorn, has been lauded to the pinnacle of bovine fame. The precursors of the herds of wild cattle of the pampas crossed the isthmus of Panama, came by ship down the west coast to Peru, then followed the old Inca road high in the Andes for two thousand miles before the hardiest reached the northern plains of Argentina. And a century later the ancestors of the Longhorn moved off the central plateau of Mexico into the chaparral of South Texas. The preservation of this nostalgic beast—which had picked up a few genes of British breeds after crossing the Rio Grande—has been assured by herds maintained in government parks and by a few enthusiastic private breeders. The Criollo founded many more cattle empires than those dominated by the Longhorn for a few decades. Yet the Criollo is passing from the scene unheralded.

The end of the Criollo tale—what can be seen today—is based on personal observations. And my observations and, wherever possible, conclusions have been substantiated by discussion with recognized authorities.

In putting together this story of Spanish cattle, I have tried to use only those particulars from historical records which are corroborated by more than one source, or which can be considered tenable in light of the circumstances involved. I have omitted what were considered to be erroneous or unconfirmed materials, unless such disclosures are mentioned to lay a fabled concept to rest. Where assumptions appear to be justified, I have indicated that they are such.

Place names in the text are those now in common usage, if the old names can be directly related to the current ones. It is surprising how many geographic names from Spanish colonial days carry down to present times. The old name is used if there is no modern equivalent.

All of the photographs were taken by my wife or by me. The thousands of pictures taken over the past decade for my books on *World Cattle* were one source of pictures. In 1975 I made another study of the cattle in Spain, the Canary Islands, Venezuela, Mexico, and Colombia (because of the special importance of these areas in the story of the Criollo) and took more pictures. From these two sources the pictures for this book were chosen, selected to show those animals most representative of their kind.

References cited in the text refer to the authors and their works in the Sources Consulted, which appears in the back of this book.

Two appendixes were added to complete the outline of the introduction of all types of cattle which reached the Western Hemisphere. Appendix I covers the introduction of the northern European and Indian Zebu breeds. Appendix II summarizes the cattle movements to Brazil.

Two old friends were helpful in putting this book together: H. H. Stonaker, formerly Dean of Agriculture at Colorado State University, and widely traveled in Latin America, read the manuscript at various stages of completion and made many valuable suggestions; and Charles R. Koch, of Oxford, Ohio, the well-known livestock editor and author, edited the final draft. Carlos Perez de Rubin y Elder of Madrid assisted most generously in the research of historical records on Spanish cattle. Professor Charles Julian Bishko of the University of Virginia, through his searching questions and pertinent comment in our extended correspondence, helped greatly in the final stages of refining the finished manuscript. My wife, Roma, again has served as photographer, researcher, and critic in the compilation of this work. As with my other writings on cattle, it is as much her book as mine.

Saratoga, Wyoming   John E. Rouse
January 7, 1977

CONTENTS

ILLUSTRATIONS AND MAPS

THE CRIOLLO

Spanish Cattle in the Americas

INTRODUCTION

The Criollo is not a breed but can be considered a New World landrace from which many types of cattle and some breeds were developed. The name designates all cattle whose ancestry traces to a few hundred head of Spanish cattle that were carried to the island of Hispaniola during the two decades after Columbus landed. Descendants of these first cattle in the New World soon spread to the other islands of the Indies, then to the mainland shores, and finally to the far reaches of both continents of the Western Hemisphere. The sons and daughters of the Spanish cow drew the plow of both peon and planter; they were the source of essential foods; and their hides provided for innumerable needs. After three centuries of adaptation to whatever environments they were trailed, these progeny of the old Spanish cattle became the *Criollo*—cattle of the country.

For the first century after their arrival the Spanish cattle were the only domesticated bovines in the Western Hemisphere, excepting their close relatives that had reached Brazil from Portugal. English and French colonists on the eastern seaboard of North America brought over northern European cattle in the early seventeenth century, but these had no contact with the Spanish cattle for another two hundred years.

In the United States the cattle from the north of Europe

came to be known as Native American. Interbreeding of the Criollo and the Native American began around 1800, first in Louisiana. Next came the herdbook cattle from England and interbreeding increased rapidly. The Zebu from India was brought to Jamaica in the 1860's, soon reached the continents, and was crossed with the Criollo throughout the tropical regions. Breeding out of the Criollo gained increased momentum with the introduction of the humped cattle.

All over Latin America, the progeny which resulted from the interbreeding of the Criollo and the new breeds of cattle were also called criollo. The emphasis of the cattleman was all on the foreign breeds—the Shorthorn, the Hereford, the Zebu, and later many others. Except in some areas of Colombia and Cuba and a few isolated communities in other countries, the Zebu- or Shorthorn-Criollo crossbred heifer was considered more desirable property than her dam. To give the pure Spanish descendant her due, she will here be designated as Criollo (with a capital C) and her hybrid offspring as criollo (with a lower case c).

The time is now 1800. Segregated by geographical features and distances, Criollo cattle exist in all inhabited areas of both continents except eastern North America.

Natural selection, under widely varied environments, and the attention man gave to breeding, have resulted in marked differences in the many Criollo populations that have covered the hemisphere. Body size, conformation, milking ability, and horn shape are characteristics that display marked variations in the cattle of different regions. Hair color patterns, however, have remained remarkably consistent.

When the cattleman in the tropics saw that he got a larger calf by putting a Zebu bull on his Criollo cows, he lost his perspective. Entranced with the Zebu he directed his breeding program to the elimination of the Criollo, an objective easily accomplished in a few generations. The fact that it was the *first* cross of the Zebu-Criollo which produced a better beef animal than the upgraded Zebu steer went unnoticed. The concept of hybrid vigor—the breeding of one pure type to a different type to obtain offspring superior to either of the parents—was unrecognized. The Criollo was bred out under the misapprehension that if a little salt was good, a lot was better.

For future cattle generations the loss of the Criollo could have even greater implications. The survival of the fittest over four centuries of natural selection produced Criollo types ideally adapted to environments that were most inhospitable to cattle. Forced to an existence in such regions as the llanos of South America, the palmetto lands of Florida, or the sierras of the Andes, strains of completely adapted Criollo evolved. They could outproduce their cousins imported from temperate climates and needed none of the pampering and clinical care required by those strangers.

The contribution that these highly differentiated gene pools might have made to the cattle of a future day will never be known. The Criollo of the llanos and the sierras are either gone or are fast disappearing. The Milking Criollo is down to a few thousand head and her future is far from assured.

The total number of Criollo cattle now remaining in the Western Hemisphere is only a few hundred thousand in a total cattle population of 320,000,000 (Brazil excluded). The part of the Criollo cow in propagating the bovine foundation on which western civilization rests may be visualized from Figure 1. This is the outline of the story to be unfolded in the following pages.

Schematic Diagram

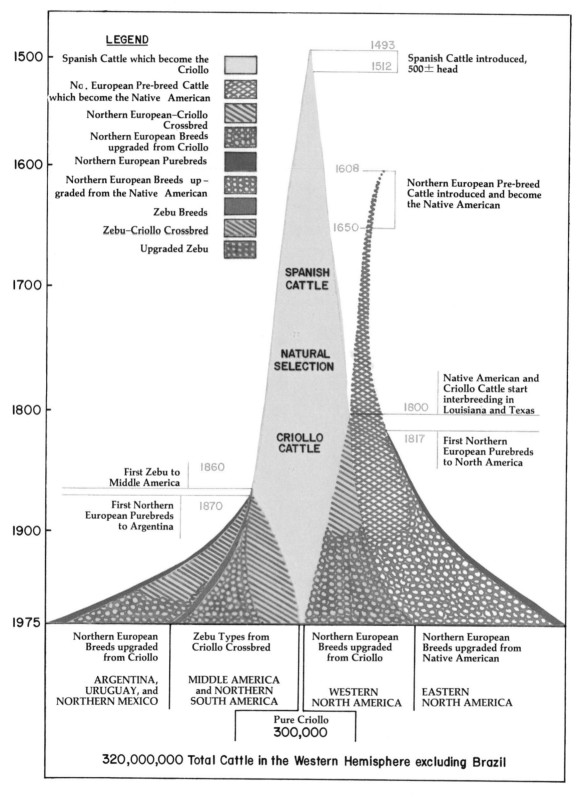

Fig. 1. The Criollo Role in Evolution of Western Hemisphere Cattle Populations

1493: The first cattle arrive in the Western Hemisphere with Columbus. The next two decades see the foundation herd of Spanish cattle established on the Island of Hispaniola (now occupied by the Dominican Republic and Haiti).

1512: The Spanish conquest of the New World proceeds. Cattle are carried to Puerto Rico, Jamaica, and Cuba, then to the mainland.

1600: Spanish cattle have penetrated both continents—south to the northern pampas of Argentina, north across the Rio Grande.

1608: Northern European pre-breed cattle reach the east coast of North America. Importations continue for forty years when the foundation herd in the English colonies becomes self-perpetuating. Native American cattle emerge.

1650: Colonial expansion of Spain in Latin America proceeds. Spanish cattle increase and wild cattle exist in uninhabited areas. The Spanish cattle become diversified into segregated populations and, through natural selection, adapt to varied climates and become known as Criollo.

1800: First contact between Criollo and Native American cattle. Interbreeding begins. Native American–Criollo crossbred appears.

1817: Northern European pure breeds reach the United States. Interbreeding with Native American begins. Native American disappears and is replaced by upgraded Northern European breeds. Purebred herds of Northern European continue as a small part of the total cattle population.

1860: Zebu breeds brought to the tropics. Zebu-Criollo crossbred appears. The Criollo influence is lost in some areas; Zebu-Criollo crossbred becomes upgraded Zebu. In other areas, Zebu-Criollo crossbred remains. Purebred Zebu herds continue as a small part of the population in Criollo areas.

1870: Northern European pure breeds reach Argentina. Interbreeding with Criollo begins. Northern European–Criollo crossbred appears, increases rapidly, and becomes the dominant part of the cattle population in Argentina and Uruguay. The Criollo influence in the Northern European–Criollo crossbred is lost. The upgraded Northern European breeds emerge. Purebred herds of Northern European breeds continue as a small part of the total cattle populations.

1900: The Criollo influence is lost in North America. The cattle population of the western United States is predominantly Northern European breeds upgraded from the Criollo.

1975: Northern European and Zebu breeds and crosses cover the Western Hemisphere. Only isolated remnants of the Criollo remain.

Part I. SPANISH CATTLE REACH THE INDIES

The cradle of the cattle industry of the Americas was in the cowpens of tenth-century Castile. Here began the movement southward to the plains of Andalusia of the cattle that were the progenitors of the Criollo. The husbandry of cattle in medieval Spain, as in all of Europe, was centered around the establishments of the feudal lords, their serfs, and the ecclesiastics. The ox and cow drew the plows and carts. The hide from everything that was killed or died went to the local craftsman. The food of the populace was largely cereals, supplemented by dairy products derived from the large flocks of sheep and goats. Beef was a luxury, mostly for the gentry, unless an animal was no longer useful. The greatest service of the bovine was to furnish energy before the plow, and to this end its husbandry was directed. All livestock were kept in close confinement.

Such were the methods of cattle management in northern Spain, to which the Moors had pushed most of the Iberian rulers by the end of the tenth century. In the succession of conquerors of the Iberian Peninsula, from the Romans to the Visigoths, the Moors were the last to hold Spain in subjection. They crossed from Africa near the Strait of Gibraltar in 711 and moved steadily northward until stopped by the reconquest in Spain which began three centuries later.

Genesis of the Criollo

Nothing is known of the physical characteristics of the Spanish cattle at this time—they were simply the descendants of old landraces which had been domesticated for centuries. The aurochs of northern Europe are known to have penetrated into the Iberian Peninsula, and before the time of Christ the Celts brought in domesticated cattle that had originated in Asia Minor. Interbreeding of these two ancient stocks produced the Peninsular bovine.

Herds of cattle were small, numbering in the tens for the large estates and monasteries to only a span of oxen for the serf. The only large herds of animals were sheep, held in flocks of several thousand by the Castilian nobles. Goats, in small numbers, contributed their share of milk and cheese.

The pages of printed history are mostly blank on cattle until the middle of the nineteenth century, even though the Spaniards were meticulous record keepers. However, paperbound tomes in the original handwriting still repose in the archives of the old walled cities of Spain, from which a wealth of information on the livestock of the Middle Ages can be gleaned. From such sources, Professor Charles Julian Bishko of the University of Virginia has uncovered the remarkable story of cattle husbandry in the Spain of the reconquest. The conclusion of the reconquest was celebrated in the year that Columbus sailed for the New World.

The following summary of the evolution of open-range ranching in Spain is based on Bishko's monograph, "The Peninsular Background of Latin American Cattle Ranching."

At the end of the eleventh century the undefined front of the reconquest was moving south, driving most of the Moorish population before it. Large areas of good grasslands were opened up south of the Duéro River and extended to the mountains of southern Old Castile. Here the running of domesticated cattle on open range began. In addition to sheep, these herds of free-roaming cattle grazed the hillsides. Townsmen, as well as the nobles and the church, became cattle owners, although sheep continued to be the preferred livestock of the aristocracy.

At the opening of the thirteenth century, new methods in cattle management began to evolve. The Castilians were now well along on the campaign that in the next 300 years was to drive the Moorish rulers out of Spain. In the wake of hostilities, as the intermittent fighting spread from the central mountains to the Mediterranean, the proud Castilian, an inveterate sheep owner, was to become the first range cattleman.

The reconquest had advanced into Andalusia by the middle of the thirteenth century. Cattle began to take precedence over sheep. The royal apportionments of grazing lands in the Guadalquivir Valley went to cattlemen instead of flock owners. The art of handling cattle under open range conditions by the man on horseback had been undergoing refinement for more than a century. The lower precipitation in Andalusia gave smaller grain yields than in the more humid areas to the north. Grass was a crop the cow harvested herself, and land was available so that free-roaming herds could be moved from one area to another as required by pasture conditions. Here was the birthplace of ranching as it was to be practiced when the cattle of Andalusia were carried to the Indies and on to the continents of America.

Castilian nobles and the monasteries ran large herds on land grants from the Crown. Some herds ran to 1,000 or 1,500 head. Prosperous townsmen owned up to 400 or 500 head on lands apportioned for common grazing. From a barnyard animal destined for the plow and milkshed, the Spanish cow had become a range animal raised for her flesh and hide.

As the range became more crowded local organizations of cattlemen were formed to control breeding and roundups, combat theft, settle disputes on grazing rights, and generally coordinate ranching operations. These control bodies, known as *mestas*, were patterned after the Concejo de la Mesta of the sheep-owning nobles of Old Castile which had exercised a self-imposed authority over the sheep industry for generations. These new mestas sometimes included sheep owners, but in Andalusia were dominated by the cattlemen.

By the time Columbus embarked on his voyage of discovery, the cattle industry of southern Andalusia had undergone two and a half centuries of organizational improvement and stabilization. Stock raising was on large holdings where geographical features constituted a major factor in controlling the movements

11

S P A

Northernmost Frontier of the Moors

Duero R.

First Range Cattle
Raising 1200 A.D.

O L D   C A S T I L E

● Madrid

N E W   C A S T I L E

PORTUGAL

CÁCERES

E S T R E M A D U R A

● Badajoz

BADAJOZ

Odiel R.

HUELVA

Huelva ●
Palos

Tinto R.

Córdoba ●

Guadalquivir R.

CÓRDOBA

● Sevilla

Genil R.

SEVILLA

JAÉN

A N D A L U S I A

GRANADA

ALMERÍA

Sanlúcar

A
Guadalete R.

Cádiz ●

Guadalete

MÁLAGA

CÁDIZ

CÁDIZ

◌ PALMA

LANZAROTE

TENERIFE

FUERTEVENTURA

GOMERA ◌

HIERRO
(FERRO)

GRAND CANARY

CANARY ISLANDS

1050 A.D.

Barcelona

N

Map 1.
Spain and the
Canary Islands

SPAIN

AFRICA

CANARY ISLANDS

of cattle. The rolling terrain of the Guadalquivir Valley, divided on the south by the Genil River and the hills and broken country east of Cádiz, naturally segregated local pockets of cattle. The Odiel and Tinto rivers, flowing through the hilly land of western Andalusia, also provided natural barriers to cattle movements.

The open-range pattern of Andalusian ranch operation, as thus described by Professor Bishko, was employed in the Indies as soon as herd size required it. From the islands, ranching followed to the mainland as cattle were introduced. On the large grassland regions in both North and South America, branding, the roundup, use of the rope, the man on horse, were universal practices. No record yet discovered reveals the kind of cattle the Andalusian stockman ran. What control was exercised over breeding or selection for a desired type or color is unknown. With so little evidence available, such matters are mostly conjecture.

Bishko describes the color of the cattle on the medieval Peninsula (Spain) as ". . . solid or mixed shades of white, cream, dun, yellow, and the light or medium reds and browns." Brand states that the Spanish cattle derived ". . . from two strains—the piebald, markedly feral range animal—and the ancient black cattle, commonly known as the Andalusian fighting bulls." Lydekker describes the color of the old Andalusian cattle as "usually dusky, although in some cases black and white, or even red and white." Such different viewpoints on color are typical of writings on ancient cattle. References to horn shapes and body size, when such can be found, are equally divergent. The only conclusion that can be arrived at is that what the Andalusian cattle looked like in Columbus' time has not yet been uncovered in the archives.

During their period of supremacy the Moors had cattle in Andalusia, but their holdings were minor compared to the large herds brought in by the Castilians as the reconquest advanced southward. North African cattle, antecedents of the Brown Atlas, were undoubtedly carried from Africa to Spain by the Moors. Brown Atlas seen today in Morocco, Tunisia, and Libya show no similarity to any cattle in Andalusia.

Like all Moslems the Moors preferred mutton and had little appetite for beef. Sheep and goat skins furnished their leather, and the heavier cowhide was little used. Thus, sheep were raised more generally than cattle. It is doubtful if any early Brown Atlas reached Andalusia in numbers that would have had a significant influence on the Spanish cattle that became established there.

The contention of some writers that the Andalusian cattle in the time of Columbus were descendants of cattle introduced by the Moors is unwarranted. This does not preclude the possibility that a few millenniums previously the cattle from North Africa reached Europe via the Iberian Peninsula in significant numbers. That the Brachyceros, or short-horned cattle, traveled from Egypt along the north coast of Africa and over to Spain at the dawn of history remains an acceptable theory. To what extent such a migration influenced the bovine population of Europe is unknown.

The assumption is made by some writers that white cattle brought in by the Romans contributed to the population on the Iberian Peninsula. It is well known that the Romans had cattle during their stay in Spain; the conquering legions could well have brought cattle along as part of their commissary. That breeding stock was shipped or trailed from Italy to Spain, where cattle already existed, is less probable, and whether the white cattle of Andalusia trace to the white cattle of Italy is yet to be confirmed.

The Canary Islands were the loading point for many cattle that went to the Indies. Spain had gained recognized sovereignty of the islands in 1479. Only meager attempts at settlement had been made previously by the French and Portuguese, and any introductions of cattle by them would have been nominal. The original inhabitants of the Canaries, the Guanches, were a primitive people who hunted the numerous wild hogs, goats and sheep on the islands but knew nothing of cattle.

The Spaniards, as they subdued the Guanches and began organized colonization, brought over cattle from Andalusia. This was only two decades before the New World was dis-

The Brown Atlas are seen across all of North Africa. These small survivors of an indigenous landrace have existed in the region for many centuries. Any cattle the Moors may have taken to Spain came from the same progenitors:

The cow of a village dweller grazing on common lands near Tlemcen, Algeria, 1965.

The nomad who owned this cow hobbled her to prevent straying; the scene is on the outskirts of Homs, Libya, 1965.

covered. Cattle ranching, as developed in Andalusia, prospered in the Canaries on a smaller scale and was a major factor in the economy of the new colony. Grand Canary, Gomera, Hierro (Ferro in old writings) and La Palma were the islands of major importance in the movement of animals to the Indies. These were normal ports of call for water and refitting, and cattle could conveniently be picked up en route.

The stock that went to establish the foundation herd in the Indies consisted of what could be purchased most readily in the areas near ports of departure and which met certain minimal requirements. Young cows in calf and young bulls, usually in the ratio of four to six cows to one bull, were mandatory. Sound, healthy animals would naturally have been demanded. Price was certainly a consideration as most expeditions were privately financed. The Crown, with few exceptions, granted nothing more than permission for exploration or colonization.

In the fifteenth century cattle in the ranching areas of Andalusia were frequently trailed considerable distances from grazier to market. Such movements involved organized drives, usually joined by more than one owner. But the law of Sevilla set a minimum of 400 head for drives of cattle requiring the use of pastures on the move to market. Since only a few dozen head were required to fill a consignment for even a large flotilla to the Indies, it seems obvious that cargoes were usually obtained near the loading point.

The available cattle history indicates that the modern breeds of cattle in Spain are, in general, found in the same regions where they have been husbanded for untold ages. Excluding the fighting cattle, there are three broad groupings of old Spanish cattle now in Andalusia:

Red: The *Retinto*, a red- to tan-colored animal sometimes almost brown. (See page 216)

Black: The *Black Andalusian*, a solid black. (See page 224)

White: The *Berrenda*, always with black markings (see page 217) and the all white *Cacereño* (see page 219). Both breeds were native to Estremadura which borders Andalusia on the northwest.

All three groupings have horns that are typical of Andalusian cattle generally—large, widespread, upturned, sometimes with a lyre twist at the ends. The hair color is solid, or nearly so, over the entire body. A high tail stock and narrow head are characteristic. These are the direct descendants of the cattle which begat those that accompanied the Spaniards to the Indies.

The old writings fail to define the kind of cattle that were shipped to the Indies, other than as to size and sex or an odd reference to a "special" type of cattle. The evidence presented by the cattle types in Andalusia today, and their counterparts seen in the Western Hemisphere, is conclusive that Red cattle, Black cattle, and White cattle with black ears were taken to the Indies. Whether the all white Cacereño accompanied them is not as definite.

Nearly all-white cattle are occasionally seen in the mountains of Central America in herds whose isolation would indicate that they are the descendants of the old Spanish cattle. On the other hand, the Cacereño was native to Estremadura and not readily available to the ports in Andalusia where cargoes originated. These white cattle are mentioned only because there is a possibility that some of them could have been included in the shipments that established the Spanish cattle in the New World.

According to some authors, the fighting cattle of Spain, the De Lidia, were man-raised in fifteenth-century Spain, but the record is not clear. Before ranching reached the Andalusian plains, bulls from the wild herds were fought or baited. The Moors, who made their contribution to the spectacle of the bullfight, are sometimes credited with bringing an old line of Egyptian fighting stock to Spain. If so, those animals would have been carefully maintained in walled enclosures and not allowed to mix with the cattle of the country. Blood-typing work of Kidd *(Breed Relationships)* has shown a significant genetic influence common to the modern Retinto breed and the fighting cattle. Similarity of genes, however, does not necessarily prescribe similarity in all characteristics of cattle in related blood groups. The African tribesman has produced remarkable differences in horn size and shape, also in hair color patterns, on cattle from the same gene pool with only rudimentary selection practices. The genetic tie between the fighting cattle and the common cattle could well trace back to the aurochs, which the early Iberians fought in the forests of Spain before the Celts arrived with their domesticated cattle.

The bullfight was an institution, and progenitors of the De Lidia were selected for aggressiveness. They would have differed widely in this trait from the old red cattle of Spain, the progenitors of the Retinto. The fighting animals have heavy forequarters, a narrow rump, and are well cut up in the flank, and in conformation bear no resemblance to the Retinto which lacks the heavy shoulders, has a fairly well-rounded rump, and a level middle. There is a marked difference in horn shape of the two breeds, particularly on the female. The De Lidia cow has forward thrust horns with a moderate upturn toward the ends. The Retinto has much larger horns, widespread and with a distinct upturn. The color of the Retinto is invariably red or dun or brown, a narrow color pattern, and always solid over the body. The De Lidia, which has an older written record than any breed of cattle, is known to have been quite varied in color—black, grey, red, various combinations of these colors, and even brindle. There was certainly a wide divergence in color, as well as in other characteristics, between the Retinto and the fighting stock as far back as there is any evidence.

There is record of "12 pairs of bulls and cows . . . the oldest fighting bull stock in Mexico" being taken to New Spain (Mexico) in 1552. This was evidently a very special cattle shipment, arriving as it did four decades after the Indies had received the base foundation stock from which the Western Hemisphere was populated. By 1552 the cattle population of New Spain must have reached at least a few hundred thousand—Coronado had entered what is now United States territory twelve years earlier with 500 head. It is obvious that 24 head of De Lidia cattle, or even several shipments of this size, all of which would have been carefully segregated from other cattle, could not have had a significant influence on the national herd of New Spain. (See page 55 for discussion of the first fighting cattle brought to New Spain.)

The cattle shipped to the Indies were to found a base of breeding stock for draft, meat, and milk animals. Handling livestock aboard the sailing vessels of the time, as well as loading and off-loading them, was enough of a chore with the ordinary farm beast that was accustomed to some control by its master. In the early days of settlement, when the islands were being stocked with cattle, there was continual trouble with animals straying and becoming wild because of the lack of enclosures to retain them. Fighting stock require stoutly fenced pastures so that animals can be controlled for breeding and maintained with a minimal presence of man. The breed is useless for draft or milk production—two objectives in the early stocking of the Indies. Such considerations would rule out the inclusion of the fighting bull or his dam in the foundation stock sent to the New World in the sixteenth century. The place allotted the fighting bull in some of the colorful writings on early cattle in the Americas cannot be accepted.

Introduction of Spanish Cattle to Hispaniola

The first cattle in the Western Hemisphere arrived off the north coast of Hispaniola in November, 1493, with Christopher Columbus, Admiral of the Ocean Sea and Governor of the Indies. The voyage was the Admiral's second to the New World and was undertaken to initiate colonization. His command consisted of 17 ships carrying 1,200 men, crew and colonists included.

Hispaniola, the site of the first colony and originally named La Isla Española by Columbus, was generally called Santo Domingo in early writings. The settlement to which the

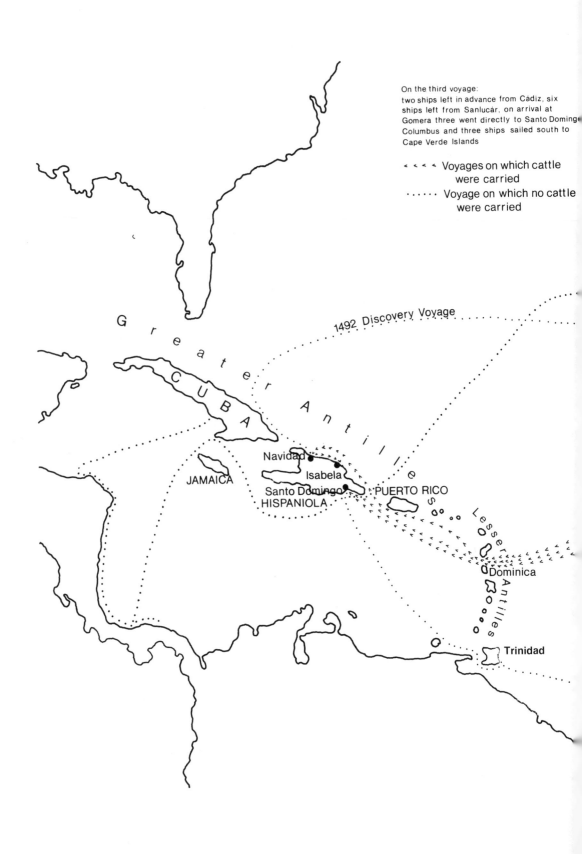

On the third voyage:
two ships left in advance from Cádiz, six
ships left from Sanlúcar, on arrival at
Gomera three went directly to Santo Domingo
Columbus and three ships sailed south to
Cape Verde Islands

‹ ‹ ‹ ‹ Voyages on which cattle
         were carried
· · · · · Voyage on which no cattle
         were carried

1492 Discovery Voyage

G r e a t e r   A n t i l l e s

C U B A

Navidad

JAMAICA

Isabela

Santo Domingo

PUERTO RICO

HISPANIOLA

Lesser Antilles

Dominica

Trinidad

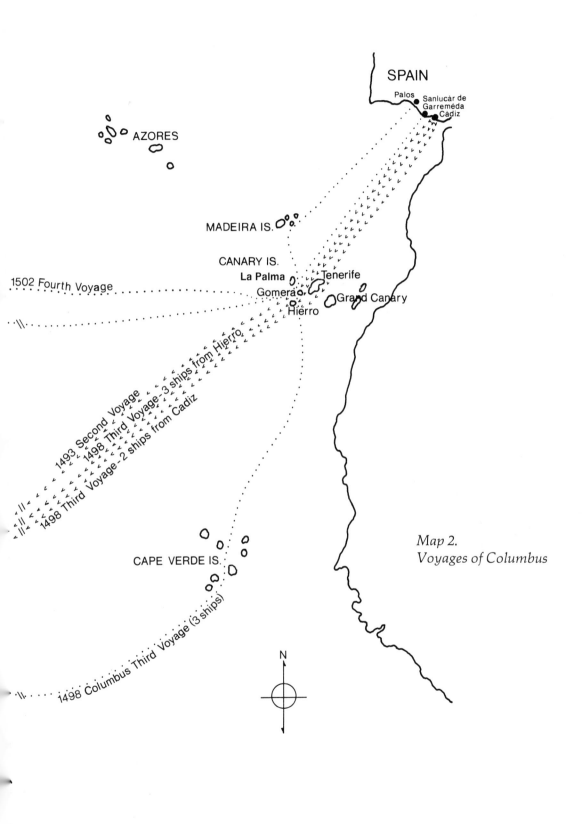

SPAIN

Palos

Sanlucár de
Garreméda
Cádiz

AZORES

MADEIRA IS.

CANARY IS.

La Palma   Tenerife

1502 Fourth Voyage

Gomera   Grand Canary

Hierro

1493 Second Voyage

1498 Third Voyage—3 ships from Hierro

1498 Third Voyage—2 ships from Cadiz

CAPE VERDE IS.

Map 2.
Voyages of Columbus

1498 Columbus Third Voyage (3 ships)

N

17 ships carried a varied cargo was intended to be self-supporting for an indefinite period. In addition to the cattle, there were horses, hogs, and sheep; there were plants, seeds, food stocks, clothing, and armaments—supplies and equipment designed to meet the requirements of the new colony and protect it in an unknown and possibly hostile land.

The demand for space in the small ships severely limited the number of animals which could be carried. Even for a large flotilla, a few tens of head sufficed for a cargo. Some cattle had been loaded at Cádiz, the port of embarkation in Spain; others were taken on when moorings for provisioning were made in the Canary Islands. History has failed to record the number shipped from either port.

This crossing to the Indies was a record one, requiring only 21 days from the island of Hierro to the first sighting of the small island of Dominica. The winds held and Hispaniola was made in 22 days. Arriving at the latter island, Columbus found that the caretaker force he had left on his first voyage at the site called Navidad had been killed by the Indians. This first habitation of the Spaniards in the Indies was in ruins. Retracing his course along the coast, Columbus landed at a new site that he named Isabela. The location was on a barren strip of coast, poorly situated for a port, but the need to get both man and beast ashore was acute.

Columbus soon realized how dependent the virgin settlement was on the lowly cow. On January 30, 1494, in a letter to the King and Queen (Ferdinand and Isabel), he cited ". . . the great need we have of cattle and beasts of burden both for food and to assist the settlers in their work," and then continued: "Plantings are far less than are necessary (for the support of the colony) . . . we had so few cattle lean and weak that the utmost they could do was very little." A little later, he wrote, "We have sown a few plots of ground."

Their Majesties agreed in principle to this appeal and said they would do what they could to send more cattle. They demurred, however, when Columbus made the suggestion, evidently to reduce the cost involved, that the cattle to be sent over should be paid for by sending back as slaves some "cannibals" (Carib Indians) ". . . because they have both better mental and physical qualifications for work than the other natives." (Major, *Letters of Columbus*)

The Spaniards were the first to move large animals on long ocean voyages; the Portuguese, who were earlier in sailing to unknown lands, were traders and did not initially found self-supporting colonies. Some experience in transporting cattle by water had been gained when the Canary Islands had been stocked fifteen years before Columbus arrived in the Indies.

From the Atlantic ports on the Iberian Peninsula to the Canaries was a voyage of 900 miles, taking four to eight days in waters that had been navigated for decades and where the wind patterns were well known. Across the ocean sea from the Canaries to the Indies was 2,500 miles, a voyage that averaged about sixty days over waters with which the Spanish navigators were just becoming familiar. Providing feed, water, and care for a few head of cattle aboard ship for such a period was a major problem for a captain without cow know-how. Shipping losses were high. In prolonged calms, when water began to run short and all fodder had been consumed, the livestock went overboard. Historians describe horses being cast overboard—hence the terminology "horse latitudes," the area where calms were encountered. Whether cattle suffered the same fate or were butchered to augment an always meager commissary on shipboard, does not appear to have been reported. They were surely disposed of one way or the other.

How cattle were loaded, confined aboard ship, and off-loaded is another blank in the record. Again it is necessary to revert to the horse, which always held a greater interest for the Spaniard than the cow, in order to obtain a clue. Illustrations in old writings show a horse being loaded aboard ship by means of a hoist, the animal suspended by a wide sling under the belly. Off-loading could have been accomplished in the same manner but only after a port facility had been constructed. How off-loading was handled in early landings, when a sandy beach was the only approach, can only be sur-

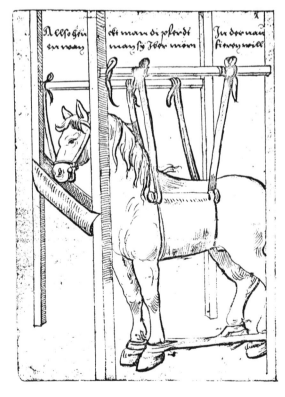

*Loading horses aboard ship in the sixteenth century. (*Manejo real en que se propone lo que deb en saber los cavalleros, *by Manuel Alvarez Ossorio y Vega, Madrid, 1769, reproduced from* The Horse of the Americas, *by Robert M. Denhardt, Norman, 1975)*

*Method of supporting horses aboard ship in the early sixteenth century. (*Das Trachtenbuch des Christoph Weiditz, *von seinen reisen nach Spanien, 1529, und den Niederlanden, 1531–32.)*

mised. Forcing an animal overboard, with a rope around the neck to entice it to shore, may have been the answer. Where the quay or wharf was available, a ramp could have been used for getting livestock either on or off board. No description of such procedures has come down to us.

Once aboard, an animal had to be confined by some positive means. Neither man nor beast could maintain footing on the small sailing vessels in a heavy sea without a fixed means of support. A four-legged animal if thrown off its feet by the pitching of a ship would sooner or later suffer broken bones and have to be disposed of. Horses were suspended by a hammock arrangement under the belly, the ends tied to an overhead support, the feet hobbled. This contrivance could be adjusted so that only the hind feet could touch the deck in rough weather. It seems logical that a similar method was employed in handling cattle. Heifers and young bulls were normally transplanted overseas, and their smaller size would simplify shipping problems to some extent. Still, a positive means of confinement and support would have been necessary.

The Spanish ships in Columbus' day carried boats for handling cargo where there were no port facilities. These were adequate for most types of goods, but it is difficult to imagine how either horse or young cow could have been transferred from a ship's deck to a small boat alongside. How these obvious problems were solved in shipping cattle during the days of Spanish conquest remains an open question. All that is known is that cattle did get across the ocean and thrived after reaching the Indies.

The ports of Spain from which ships sailed for the Indies were Sevilla and sometimes Córdoba, both on the Guadalquivir River; and Palos, Sanlúcar de Barremada, and Cádiz on the coast. During the early days of colonization, Sevilla was usually the primary port of departure, and cattle were certainly loaded there. The course was then down the Guadalquivir River to the coast and on to the Canaries. Palos, occasionally a port of departure, could have loaded cattle that came from the valleys of the Tinto and Odiel rivers. Any cattle taken on at Cádiz would have originated in the rolling country to the east. Small vessels leaving from Córdoba, one hundred miles up the Guadalquivir River from Sevilla, also loaded cattle for the New World. On the routine call at the Canaries, the cattle taken on came from the islands where moorings were made.

The movement of livestock was greatly facilitated by loading at the Canaries rather than at a mainland port. Because the time at sea was reduced one-fourth, shipping losses were fewer, and the animals arrived in the Indies in better condition than those leaving from Spain. These factors were noted by the cattlemen of the New World. The frequent requests to the Queen and King for more cattle often asked that shipment be made from the Canaries. When the *Casa de Contratución* (House of Trade) was established in 1503 to direct and organize all shipping to the Indies, the Canaries were often specified as the loading point.

The largest vessel in any of the early fleets to the Indies was described as of 150 tons burden; the average was nearer 100 tons. Columbus' flagship, the Santa María, had a deck length of about 85 feet, a beam of 25 feet, and a rating of 120 tons. The necessities of nature were accomplished by hanging out on the rigging. The requirement that young bulls and pregnant heifers be selected for a cargo was to keep the weight down and for greater ease in handling as well as to get "two heads in one." Even on a bare maintenance diet the animals required a quantity of feed and water three times their own weight for a sixty-day voyage.

The orders issued pursuant to Columbus' request in 1494 to the Queen for more supplies for the Isabela settlement specified: "6 mares, 4 donkeys and 2 females, 4 heifers and 2 young bulls, 100 sheep." (Mesa, *Historia de las Primeras Importaciónes*) Hard pressed as the struggling settlement was for cattle it was allowed only six head. Additional cattle reached Hispaniola during the years that elapsed between 1493 and the arrival of the first two vessels of Columbus' third expedition in 1498. The number was certainly recorded in the manifests of early shipping that once reposed in the Archives of the

Indies in Sevilla, but, apparently, did not attract the attention of historians.

The Admiral had only eight ships under his command for his third voyage. Before leaving Sanlúcar in May 1498 he dispatched two vessels from Spain to Hispaniola. When Columbus with six ships reached the Canaries, three were sent direct to the colony. In total, five vessels of this voyage probably carried cattle, though the number actually off-loaded must have been small.

Columbus was not in high favor with the Court at the time. In the hope of a discovery that might restore his position, he took off on an exploratory voyage with three ships and discovered the island of Trinidad. He reached Hispaniola in August, 1498, his vessels carrying no cattle. During the Admiral's absence, no progress had been made in establishing an adequate breeding herd on Hispaniola. The state of the island's cattle, however, was the least of the disappointments Columbus encountered on his return. An armed revolt was in progress led by one Francisco Roldán, the man Columbus had appointed mayor of Isabela at the time of his departure. The town had been practically abandoned; and a new port, founded in 1496 at Santo Domingo, became the island's principal city.

Roldán had seized the opportunity to foment a rebellion among a dissatisfied element of the colonists while Columbus was away. News of this disruption of authority, along with many reports of mismanagement by the Admiral, had reached Spain. The Queen had then sent Francisco Bobadilla, a knight with no background of administration, to investigate conditions and restore order. Bobadilla quickly assumed command of the island and was recognized as governor. Columbus, still Admiral of the Ocean Sea but no longer Governor of the Indies, and his brother, who had remained on Hispaniola, were soon in chains and on the way back to Spain. Bobadilla, after a fashion, brought law and order to the island, but his administration was wantonly corrupt. The shortage of cattle continued.

Once back in Spain, Columbus gained the ear of the Queen, convinced her he had suffered unjust treatment at the hand of Bobadilla and regained to some degree the royal favor. Bobadilla was recalled, and by royal decree Frey Nicolas de Ovando was named governor in 1502. Ovando, a military man, was held in high regard at Court and had shown ability and integrity in his commands.

The low point in Spain's conquest of the Indies coincided with the recall of Bobadilla. Hardship, disease, even famine at times, had decimated the colony. Disillusioned repatriates returned to Spain whenever they could arrange for passage. Only 300 Spaniards remained of nearly 2,100 who had arrived up to this time. The island's cattle herd was in sad disarray; the number of wild animals probably exceeded those that remained confined. Breeding as well as draft animals had been consumed for food. The shortage of cattle was a major factor in the disintegration of the colony.

Ovando arrived in April, 1502, at Santo Domingo. He commanded the largest flotilla to reach the Indies up to this time—30 large ships and 2,500 men, with livestock and all the armament and supplies needed to revive the colony. A man of action, he rapidly subdued the Indian population, often by treacherous slaughter of the caciques (Indian headmen) and their immediate attendants. Ovando's methods, although needlessly brutal, soon brought the natives under control.

For several years after Ovando's arrival, the shortage of cattle continued in spite of the additions that his expedition had contributed. One message he sent to the Queen soon after he landed stated, "Santo Domingo (Hispaniola) must have an abundance of cattle to avoid trouble . . .," and also ". . . production of meat is the most urgent need of all the people." (Puente y Oleo, *Los Trabajos Geográphicas*) This brought a strong response and cattle shipments, along with other livestock, began to arrive.

At this point, a new authority entered the administrative scene in Spain which did much to stabilize the efforts at colonization and conquest of the New World. *La Casa de Contratación* was established by royal decree in 1503 to authorize and direct all commerce and shipping, particularly in matters involving the Indies. The *Casa* was all powerful and did much to organize the shipping and trade that was essential to orderly colonization.

Vicente Yáñez Pinzón and Juan de la Cosa, both of whom had been with Columbus on his first voyage, were instrumental in the formation of the *Casa*. They were influential gentlemen with wide experience in the Indies and firsthand knowledge of the logistic problems involved. The establishment of breeding herds, not only on Hispaniola but later at other key points of colonization, was accomplished through orders issued by the *Casa* to transport cattle to specific destinations.

As soon as the *Casa de Contratación* was formed, the Queen ordered that authority to see that all ships going to the Indies carried cattle, as well as other animals, and, further, that all ships arriving in Spain or at the Canary Islands from the Indies take cattle on for the return voyage. Over the next few years there were frequent requests from Ovando for more cattle and repetitions of the *Casa*'s order to send more.

The severe hurricane which crossed Hispaniola in 1508 was a heartbreaking blow in the effort to establish an adequate breeding herd. "Many of the new plants, most of the already too small cattle population, and the houses of adobe and thatched roofs were destroyed." (Puente y Oleo, *Los Trabajos Geográphicas*) A further obstacle imposed by nature was the plague of insects (probably locusts) which devastated crops in Andalusia in 1508. This placed cattle in short supply in the homeland. The *Casa de Contratación* was ordered by the King to see that cattle and other supplies were shipped at once from the Canaries on every ship going to the Indies.

Hispaniola, with Santo Domingo the major port and capital, was to serve as the base of conquest and colonization in the New World for the first two decades of the sixteenth century. It naturally followed that here was the major supply point from which the rest of the Indies and the continent were to be stocked with cattle. The difficulties encountered in getting an adequate breeding herd established in Hispaniola delayed the stocking of the other islands.

The early cattleman on Hispaniola occupied a villa and obtained pasture rights to the surrounding open range. Originally the villa, located near an Indian village whose inhabitants were forced to work the placer mines, was a group of buildings housing a ranking Spaniard and his retainers. It was gold in the stream beds of Hispaniola that held the Spaniards on the island until it grew to be the operation base for the Indies. When the gold played out, the villas became New World ranches. Later villas were established solely for stock raising in areas where Indians could be found to do the work.

As far back as 1498 royal villas were established for raising cattle in order to increase the supply of breeding stock for colonization. As conditions stabilized, open range management as developed on the Andalusian plains became the hidalgo proprietor's pattern of cattle raising. Horses and hogs also ran free and were utilized as needed. The pasture rights eventually became land titles, horses were employed more extensively in working cattle, and roundups were held to take off the surplus for market. Hides often became more important than beef. A wider utilization of leather than was common in Andalusia had been introduced by the Moors. In the New World, however, the shortage of many of the ordinary articles of commerce led to the use of leather for many new purposes. As colonization spread over the hemisphere, the demand for hides increased as the general utility of leather became appreciated. It was the material for more articles of clothing, furniture, and shelter. Excellent rope was fabricated from leather. Nails were always scarce, at times being mentioned as worth their weight in gold, and rawhide proved an effective substitute. Used as a binding it served as the means of holding wood members together in many types of construction, from chairs to buildings. The importance of cowhide to the economy all through colonial days is difficult to visualize today.

The cattle industry served the island well as the transition was made from a gold-mining to an agricultural economy. Once established, cattle did well on the pastures of Hispaniola. The increase was greater than in Spain or the Canaries because of the continuous growth of green forage. The Andalusian cattle were well adapted to a hot climate, but in their homeland they had to endure the long dry winters with only weathered standing grasses for sustenance. On the year-round pastures of the island, cattle and all livestock reproduced phenomenally in both controlled and wild states.

In the early years of the second decade of the century, the cattle population of Hispaniola reached the level of surplus. Authentic reports state that cattle ran wild in 1512 and were being killed only for their hides. A few earlier writings, one even in 1507, mention large numbers of wild cattle, but these lack credibility as they are concurrent with urgent requests for more cattle and the orders from the *Casa de Contratación* to ship them. The setting of 1512 as the date when Hispaniola was amply supplied with cattle for export assumes a bit of latitude, but can be taken as a reasonable approximation. It is clear that wild herds existed on the island at this time, and if cattle were available simply by rounding them up, it is hardly conceivable that people needing cattle would undertake the expense and involved procedures required to obtain authority for their importation.

The Spanish authorities involved with shipping, particularly the *Casa de Contratación*, kept meticulous records on all matters pertaining to the Indies. A search in 1975 in the General Archives of the Indies in Sevilla for manifests of ships carrying cargoes to the New World uncovered many of the old original records, dating from 1529 to 1599. In the ships' manifests no mention was found of any type of cattle (*vacas*, cows; *toros*, bulls; *novillos*, young bulls; or *terneras*, young cows), although there were many detailed listings of horses and every conceivable article of human need.

It seems reasonable to conclude from this bit of negative evidence that very few cattle were loaded for the Indies during this period. Certainly there would have been no occasion to ship cattle if the grasslands of the New World were overrun with them. Ship manifests before 1529 could not be located in the archives, and early cattle movements could be traced   *33* only by occasional mention in contemporary writings. Such records specify actual cattle shipments, authorization for shipments in detail, and many requests for cattle up to around 1512. After that they are largely silent.

Stocking the Indies

Cattle followed the conquistadors with the ecclesiastics and settlers as they advanced from Hispaniola to Puerto Rico, Jamaica, and Cuba. These four islands to become known as the Greater Antilles—were "The Indies" in Columbus' day. Spain in her conquests ignored the small islands, the Lesser Antilles, although these also came to be included in the general term, the Indies.

The foundation herd that had been established on Hispaniola by 1512 was the principal source of the breeding stock that went to the other islands. Small vessels had been brought over from Spain for interisland shipping, and the port at Santo Domingo was the hub of commerce in the Indies. Although an occasional consignment of cattle went to one of the other islands direct from Spain or the Canaries, the main supply point was Hispaniola.

Puerto Rico   One shipment of cattle reached Puerto Rico direct from Spain before the Hispaniola herd had increased to the point where it could be drawn on for breeding stock. The first bovines appear to have arrived on the island in 1505.

Vicente Yáñez Pinzón, one of the three brothers of the influential shipping family which had participated in financing Columbus, had captained one of the ships on the discovery voyage. In 1505 he was appointed governor of Puerto Rico. The *Casa de Contratación* had issued an order to take ". . . plants, cows and other supplies to Puerto Rico for the establishment of a settlement as a base for further exploratory effort." Implementing this order, Pinzón left Palos in September with the required cargo, but on reaching Puerto Rico, "an accident [not further described] caused the loss of plants, supplies, and the *cows escaped to the mountains*." (Puente y Oleo, *Los Trabajos Geográphicas*)

There are later references to these cows multiplying and existing in a wild state. How many there were, or to what extent they may have influenced the cattle population on the island is not known. In all probability their descendants continued in a wild state and must have mixed with cattle introduced later as Puerto Rico was settled. This instance was one of the occasions where cattle from overseas went directly to a port other than Santo Domingo.

The practical foundation of the Puerto Rican breeding herd began with the shipments made by Ponce de León in 1509. Gold had been discovered on the island and, to hasten colonization, the King authorized the importation of cattle from Hispaniola by the incoming colonists. Otherwise, exports from Hispaniola were prohibited, because the shortage of cattle on the island was still critical.

Juan Ponce de León, a cavalier of an ancient Castilian house who had served in the wars with the Moors, had come to the Indies with Columbus in 1493. He became a lieutenant of Governor Ovando in the subjugation of Hispaniola and acquired a villa and a substantial cattle operation. Named governor of Puerto Rico to succeed Pinzón he left Hispaniola early in 1509 a wealthy man and well equipped to found a capital. Breeding stock from his herd was included in this move. Later in the year, even after the great hurricane had further depleted the herds on Hispaniola, de León was given permission to remove an additional "50 young bulls."

In 1511 there appears to have been a second movement of cattle to Puerto Rico. An early cattleman who had brought 36 cows and a bull to Hispaniola in 1504 was given special permission to take his herd with him when he moved to Puerto Rico. After 1512 breeding stock was readily available on His-

Tampa

Havana

C U B A

Asunción de Baracoa

Santiago Maisi

1511

Navidad Isabela

HISPANIOLA La Vega

Santo Domingo

Salvatierra de la Sabana

1509

JAMAICA

1509

N

P A N A M A

PUERTO RICO

San Juan

1493

Guadaloupe

Dominica

From
Spain
and
Canary Is.

1494-1512

SOUTH AMERICA

Map 3.
Early Inter-island
Cattle Movement in
the Indies

paniola and moved freely to the younger colony as the demand increased. By the end of the decade cattle, reported to be running wild on the island, were evidently now more numerous than could have been accounted for by the cattle that had escaped from Pinzón in 1505.

Gold was the incentive for the colonization of the island, sizable placer deposits having been located in 1508. Demands for cattle followed the rapid establishment of mining operations. Villas with large herds were founded, and the cattle industry became a substantial factor in the economy. The requirements for edible beef were soon filled, and hides and tallow became the principal products of ranching. After 1520, Puerto Rico was self-supporting in cattle.

There was one later importation, however, of particular interest, both because it came directly from Spain and also because of the reference to a special kind of cattle. In 1541 a shipment of "*selected* cows and additional bulls" was made from Sevilla to Puerto Rico. (Patiño, *Plantas y Animales en América*) The designation "selected" is significant. There was no need to go to what was an expensive undertaking to move stock from Spain at this time if the desire was simply to obtain cattle. These could be purchased from owners on the island for the price of a hide or even by rounding up wild animals in the hills. The apparent explanation for this unusual shipment is that the enterprising hidalgo of some villa wanted new stock to improve his herd, or possibly he wanted the black, or white, animals he had seen before moving to the Indies.

Jamaica   No gold had been discovered in the streams of Jamaica, and the Spaniards had given little attention to the island. Panama, however, had caught the eye of the conquistadors, and Jamaica was recognized as a steppingstone to the mainland. In 1509, Juan de Esquivel was assigned the task of subduing Jamaica by Columbus' son, Diego, who had succeeded Ovando as governor of Hispaniola. Esquivel had accompanied Columbus on the second voyage and had served in the conquest of Hispaniola. The taking of Jamaica was accomplished by a particularly brutal slaughter of Indians, but settlements were then established and agricultural development was initiated. Cattle as well as other livestock were brought from Hispaniola and did exceptionally well.

Here, again, open-range stock raising was practiced and sizable herds were developed by the more enterprising settlers. Colonization of Jamaica coincided with the establishment of a settlement on the northwest coast of South America and the subsequent movement westward into Panama. Jamaica growers for more than a decade had a good trade in supplying breeding stock to the isthmus. This was in addition to the normal business in hides. Within ten years after their introduction the cattle on Jamaica had multiplied to the point that they had become wild in areas beyond the control of their owners. After the early 1520's, there was no occasion for further importations to the island.

No cattle had reached Cuba before the expedition of Diego Velázquez de Cuellar in 1511. Velázquez had been with Columbus on his second voyage and earned his tarnished spurs in needlessly cruel encounters with the Indians on Hispaniola. Diego Columbus, as governor, issued orders to Velázquez to subdue the island of Cuba.

Cuba

The conquistadors left from Salvatierra de la Sabana on the southwest peninsula of Hispaniola and landed at Maisí on the eastern end of Cuba with a complement of cattle. Most of these early bovine arrivals probably went to the commissary, but a breeding nucleus was kept for the villa that Velázquez built at Asunción de Baracoa soon after arrival. This foundation herd was supplemented by a number of later shipments.

The early discovery of gold by Velázquez' conquistadors led to the frantic opening of placer-mining operations. Gold took preference over stock raising, and cattlemen were slow in becoming established. The first villas were organized to support the placer mines; cattle were handled in a desultory

39

manner, and many of the imports were consumed for food. As mining operations became stabilized, cattle-raising villas, later known as *"hatos,"* were organized by the more favored adventurers from Hispaniola. Land use permits were allotted by units also called *"hatos."* The *hato* was an area one league in diameter encompassing approximately 4,500 acres. One *hato* was allotted for a minimum of 2,000 head of cattle; two units, 6,000 head; and three units, 10,000 head.

Hernán Cortés, an adventurous scion of an old Castilian family that had migrated to Estremadura, was one of the early land grantees. He had been a successful cattleman on Hispaniola but left to join Velázquez in the conquest of Cuba. After the island was subdued, he reentered the cattle business and prospered to such an extent that he largely financed his future conquest of New Spain by the sale of his cattle and properties.

The better grasslands of Cuba were loosely occupied by privileged grantees—former military men and favorites of the Court in Spain. The circular units of land measurement did not limit grazing practices, and cattle were handled on an open-range basis. Cattle increased so rapidly that many wandered from the *hatos* and became wild. Within a decade Cuba was to become the main supply point for New Spain's breeding stock.

Antilles By 1520 the Greater Antilles were in a position to stock future colonies with cattle as the conquest of the New World moved onto the mainland. The advances of the conquistadors depended on an adequate supply of cattle to follow in their wake. The major conquests always moved into the settlement phase from a base of cattle. The conquerors did not move on to adjacent islands from Hispaniola until its cattle numbers were up. With cattle in good supply there, advances to Puerto Rico, Jamaica, and Cuba followed rapidly.

Cuba generally had the cattle to support the settlements founded by Cortés, although at times there were delays in meeting the demands of New Spain and cattle were obtained

40

from Hispaniola. Panama was largely supplied from Jamaica. The north coast of South America depended on Hispaniola and Puerto Rico for cattle.

Horses, and in some localities, mules, were essential for the Spanish advance, but for successful colonization a ready source of cattle was usually the first requirement.

All the islands of the Lesser Antilles were hospitable habitats for cattle. They were stocked from Puerto Rico and Hispaniola, and later from Trinidad as colonization proceeded. Though discovery of nearly all the islands was claimed by Columbus, the early Spaniards displayed little interest in them except for slave-raiding expeditions. The Bahamas (Lucayas at that time) were actually depopulated by slave raids in the early sixteenth century after the Indian slaves on Hispaniola had been killed off by the brutal way they were worked. Such raids brought no cattle, as no permanent habitations were established. Gold, a passage to the Orient, and the claim of large areas for the Crown were the stimuli that drove the Spanish conquerors. All such incentives were lacking in the Lesser Antilles. When other European nations came to fight for the crumbs in the next century, many of the islands were still mere havens for pirates and buccaneers.

The introduction of cattle to the Lesser Antilles followed the establishment of permanent settlements during the first half of the seventeenth century. The British, French, and Dutch were the colonizers, and change of possession among these nations was rather frequent. Spanish cattle were obtained from the most convenient port to meet the demands of the rapidly developing sugar economy. All the larger islands were soon well stocked with cattle, and there was no occasion to bring more from Europe. The English planters on Jamaica showed no interest in purebred cattle until the middle of the nineteenth century.

Bermuda was far off the routes of Spanish conquest to the Indies, although its sixteenth-century discovery is sometimes

Lesser Antilles

41

claimed for the Spaniard Bermúdez. A company of British colonists, shipwrecked in the area in 1609, claimed to have found remains of abandoned Spanish shelters. Wild hogs, also thought to have been left by Bermúdez, were later found on the island. Cattle were sent out from England with colonists who landed in 1612 and again in 1614. These introductions were the English pre-breed cattle. They did well, and within a few years laws were enacted requiring owners to keep their animals secured.

Only one possible introduction of Spanish cattle to Bermuda is mentioned. In 1617 some settlers on that island sent a vessel to the "savage islands [Virgin Islands] for cattle and goats" for their colony. (Wilkinson, *The Adventurers of Bermuda*) No mention is made, however, of any animals being acquired. The cattle on the island appear to have been descendants of English cattle and do not enter into the story of the Criollo.

PART II.   ADVANCE TO THE CONTINENTS

The cattle that reached the Indies from Spain and the Canary Islands could probably be counted in the low hundreds and certainly totaled fewer than 1,000 head. The natural increase from this small gene pool served to populate the four islands of the Greater Antilles. There were only incidental additions to this population of breeding stock for three and one-half centuries. Such increments were all Spanish cattle. It was this small foundation herd, held originally on Hispaniola, then expanded to Puerto Rico, Jamaica, and Cuba, that stocked the Spanish colonies on both continents of the Western Hemisphere. Uncounted millions of its descendants eventually ranged from the Argentine to Canada. In 1783, Buenos Aires alone shipped 1,400,000 hides to Europe. Several Mexican haciendas were said to be branding over 20,000 calves a year at that same time. No external cattle types were to be mixed with the progeny of the Spanish cattle until after the opening of the nineteenth century.

First Cattle on the Mainland

Having traced the trail of the Spanish cattle to Hispaniola and then to the other islands, we can now follow them onto the

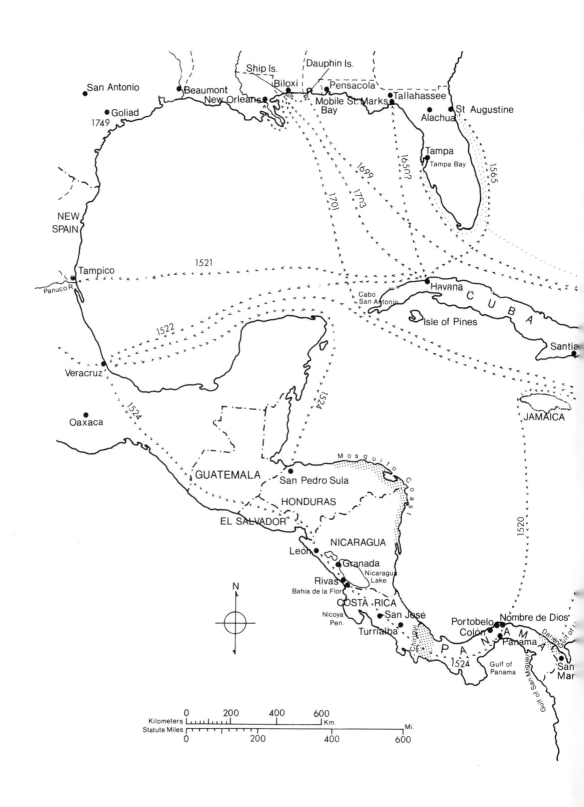

San Antonio

Beaumont
New Orleans
Ship Is.
Dauphin Is.
Biloxi
Pensacola
Mobile
Bay
St. Marks
Tallahassee
St. Augustine
Alachua

Goliad
1749

NEW
SPAIN

Tampa
Tampa Bay

1699
1650?
1565

Tampico
Panuco R.

1521

Havana
Cabo
San Antonio
CUBA
Isle of Pines
Santia

1701
1703

1522

Veracruz

1524

Oaxaca

1524

JAMAICA

Mosquito Coast

GUATEMALA

San Pedro Sula

HONDURAS

EL SALVADOR

1520

NICARAGUA
León
Granada
Nicaragua
Lake
Rivas
Bahia de la Flor

N

COSTA RICA
Nicoya
Pen.
San José
Turrialba
Chiriqui
Portobelo
Colón
Panama
1524
Gulf of
Panama
Nombre de Dios
PANAMA
Darien
Gulf of
San
Mar
Gulf of San Miguel

Kilometers
0   200   400   600
Km.
Statute Miles
0   200   400   600
Mi.

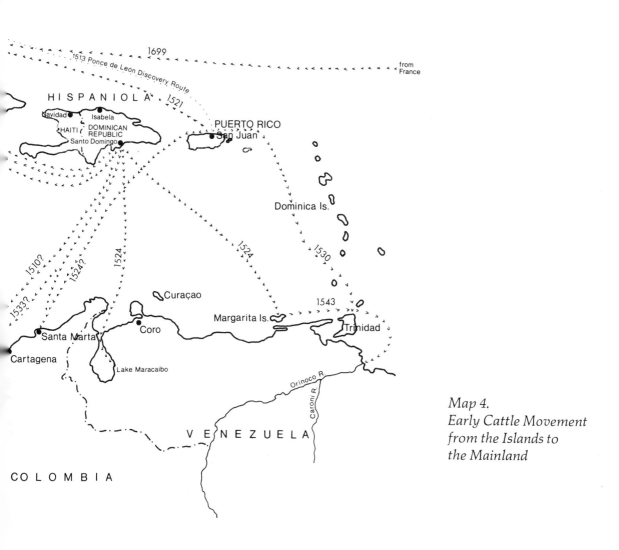

Map 4.
*Early Cattle Movement
from the Islands to
the Mainland*

mainland. The modern names of the islands define fixed areas, are directly related to those used in Columbus' time, and are easily identified with current political entities.

Moving to Terra Firma, however, becomes a complex puzzle if an attempt is made to correlate old Spanish names with specific areas or old political boundaries. When the first cattle were taken to the mainland, there were no clearly defined boundaries of Spanish authority. Later, as viceroyalties were established, colonial administration became more organized. But, again, areas of government were loosely defined. The seats of royal administration were also moved from time to time as the patterns of colonization and trade changed. When the rule of Spain faltered and the wave of independence overran the hemisphere at the opening of the nineteenth century, the present-day political boundaries began to take form.

To attempt to relate the cattle movements throughout the old Spanish Empire to the changing lines of territorial authority and geographic divisions is beyond the scope of this story of the Criollo. Thus, the names of present-day political entities will in general be used in narrating the introduction and advance of Spanish cattle on the continents.

The first breeding herd on the mainland was established around 1510 on the eastern end of the Isthmus of Panama. New Spain was next, the first cattle arriving there in 1521. The north coast of South America began getting cattle around 1524, and the west coast ten years later. The first cattle to persist as a breeding herd in territory that is now the United States were the Spanish cattle taken to Florida in 1565; these were followed by those that went to New Mexico in 1598. The French got a few cattle from Hispaniola in 1701 for the party that was to found New Orleans. Texas soil did not have domesticated Spanish cattle until 1717, although wild cattle had certainly crossed the Rio Grande much earlier. The first cattle in California arrived in 1769.

Panama and New Spain were adequately supplied with cattle after some early shortages. There was considerable delay, however, in getting animals to the settlements on the north coast of South America as rapidly as the inhabitants asked for them. As colonization progressed in territory that is now the southern United States, cattle were readily obtainable except in Louisiana. The Spaniards were exceedingly reluctant to furnish livestock to the French around New Orleans. They viewed these colonists as occupying land that should have belonged to Spain.

Brazil has been omitted from this story of the Criollo, as the country was stocked with cattle from Portugal. However, such evidence as is available indicates that the Portuguese and Spanish cattle at the time of their introduction to the Western Hemisphere were practically identical. The same range-type management was common to both countries. A resumé on the introduction of cattle to Brazil and their progress is given in the Appendix.

Panama, Nicaragua, Costa Rica

In the opening years of the sixteenth century early attempts to gain a foothold on the north coast of South America had been repeatedly driven off by hostile Indians. An expedition from Hispaniola finally secured a landing in 1510 on the Gulf of Urabá where the Isthmus of Panama joins the continent. Forced by the natives to abandon the east shore, the Spaniards transferred their headquarters to the Indian land of Darién on the west shore. Here Santa María (Santa María la Antigua de Darién) was built. It adjoined an Indian village, also called Darién, that was inhabited by a peaceful tribe which proved quite hospitable. Santa María was the first permanent Spanish establishment on the mainland and served as the base for the discovery of the Pacific Ocean and for the colonization of Panama and westward as far as Nicaragua.

The record is not clear on the date the first cattle reached Darién. The difficulties encountered by the founding expedition would certainly have resulted in any livestock it carried being either lost or consumed as food. But supporting vessels must have brought cattle soon, for in 1513 the inhabitants complained that ". . . fierce tigers killed cows, young cattle, and

47

even men." In 1515 reports mentioned wooden carts drawn by cattle. (Patiño, *Plantas y Animales en América*) Whatever the exact date of arrival, these were the first cattle on the continent.

Pedro Arias de Ávila, who became known in history as Pedrarias, arrived from Spain as governor of Santa María in 1514. A military officer and of the nobility, he proved to be a highly incompetent administrator. The march of events, however, indicate that he was a lover of cattle and realized their value in colonization. After becoming governor he probably built his herd from cattle already in Darién. The armada he commanded carried no provisions, for the adventurers who accompanied him had agreed to live off the land on arrival. When in 1519 Santa María was abandoned, Pedrarias moved his herd to the new settlement of Panama on the Pacific side of the isthmus.

Panama received stock for beef and breeding from both Jamaica and Hispaniola but was short of cattle for the first few years after settlement. There is a record of two oxen being shipped to Panama in 1520 from Jamaica. A year later, authorization was granted for the importation of "50 cows and 50 young bulls, 200 sheep and 1,000 hogs from Jamaica." Some of the bulls were evidently destined for slaughter. Shipments then increased, and by 1527 Panama was well stocked with cattle. A heifer at this time sold for 2 or 3 pesos. (Patiño, *Plantas y Animales en América*) When the conquistadors were leaving to conquer Peru in the 1530's, there was no difficulty in supplying them with beef. Large estancias, a few with herds of two or three thousand head, were flourishing from the vicinity of Panama City as far north as the Chiriqui area by the end of the sixteenth century.

Central America to the west of Panama received its livestock mainly from New Spain, but Costa Rica and southern Nicaragua were originally stocked from Panama. Two early settlements were started around Lake Nicaragua in 1524 which may have had a few cattle as well as other livestock. The region had been conquered in that year by Sebastian de Belalcazar, a lieutenant of Pedrarias, who ten years later was to rise to

48

greater fame in Ecuador and Colombia. The first recorded movement of cattle to the area was by Pedrarias, who after his failure in Panama, had been made governor of virgin Nicaragua in 1526. The herd he had moved to Panama from Darién nine years previously, he then took to Nicaragua. Ten years later there is record of another authorization for a herd to be moved there from Panama.

Since gold discoveries were lacking at the time, the sixteenth-century development in Nicaragua had to depend on agriculture even though there was little market for the products. By the end of the century, however, large herds of cattle were being run on the Pacific coastal plains. The lowlands on the Caribbean side, known as the Mosquito Coast, were soon found to be too wet for ranching. Although the initial settlements in Nicaragua, as well as the introduction of cattle, had been made from Panama, the foundation stock in the better grazing areas to the west later came from New Spain through Guatemala.

Costa Rica, although bordering Panama, was colonized later than Nicaragua. Cattle were trailed down from the Lake Nicaragua area after the middle of the sixteenth century, a few head arriving in 1561. A thousand cows were brought to Costa Rica in 1573, and in that same year there is record of 2,000 head being purchased for 5,000 pesos for future delivery. (Patiño, *Plantas y Animales en América*) The basic breeding herd in Costa Rica was well established before the end of the century.

The Land of Cortés

The flotilla with which Cortés began his conquest of the region that is now Mexico probably carried no cattle, contrary to some reports. Nor had the two earlier expeditions which had reconnoitered this part of the mainland introduced any livestock. Cortés had only 14 horses when he landed at the site of Veracruz in 1519 to establish his base of operations. The cavalry was the most prized arm of his fighting force, and if his ships could carry only 14 horses, there certainly was no space for cows.

Neither is it likely that the large expedition which arrived a year later brought any cattle. This was a military force sent by Velázquez, governor of Cuba, to take Cortés into custody and return him to Cuba. Any livestock carried would have been for the commissary. Here Cortés displayed his fantastic ability to turn adversity to his favor. He subdued the forces sent to take him prisoner and then enlisted their support in his advance on Tenochtitlán, the Aztec capital which later became Mexico City.

New Spain: Mexico, Guatemala, El Salvador, Honduras

The first cattle to found a breeding herd in New Spain probably landed in 1521 on the banks of the Pánuco River near the site of Tampico. The expedition was led by Gregorio Villalobos and was the first effort at colonization. Villalobos was known as a leader with ability and was later given recognition by being named lieutenant governor of New Spain. The point of origin of the cattle is in dispute, both Hispaniola and Cuba being named by different writers. The number is sometimes given as 50 head, and they are described as being calves. The main movements of cattle that founded the breeding herd in New Spain were unloaded at Veracruz. Shipments were made from Havana as the nearest source of supply until Cuba began to complain about a shortage of cattle. Santo Domingo then became the supply point temporarily.

The pages of history are filled with the details of valor and achievements of the conquest, but there is scant mention of the cattle that supported the missions and eventually stocked the ranges of New Spain. When once established, cattle multiplied as rapidly as they had throughout the Indies. By the time the Castilian cavaliers, Crown officials, and ecclesiastics began to arrive from Spain, there was no need for bringing in additional livestock.

Cortés had subdued the major Indian nations throughout

the territory, although a few hostile local tribes had to be dealt with later. Settlement followed in the wake of conquest. The early colonizers came from both Cuba and Hispaniola. Both the adventurous and the disillusioned, of which there were many, were eager to seek their fortune in New Spain. The leaders among the colonizers, though, were favored conquistadors and the nobility of Spain, particularly the Castilians who had become ranchers in Andalusia. They brought their open-range methods of cattle management to the virgin grasslands when they established their haciendas. The valleys and coastal plains south of Mexico City were stocked first, then the high plateau region to the north.

Guatemala and Honduras were also quickly overcome by Cortés. Colonization proceeded slowly but in an orderly manner in these areas. The route of the colonists led southeastward from Veracruz and Oaxaca, then through the gap in the main cordillera and onto the coastal plains of the Pacific.

The movement of settlers with their cattle to Guatemala began soon after the region had been conquered in 1524. Cattle were also sent to the San Pedro Sula port on the Caribbean coast of Honduras by ship from Veracruz, but here they were generally isolated. The ranching operations were mostly confined to the better cattle country on the Pacific coastal plains.

A brand book was established in Mexico City in 1529, and all owners of cattle were required to register their brands. Legend has it that the three-cross brand of Cortés was the first one recorded. A national livestock association, patterned after the local mestas in Andalusia, was founded in 1537. All hidalgos with cattle belonged to this organization. The conquistadors, accompanied by the emissaries of the church and followed by the cattlemen, paved the way for the colonization which spread northward to the Rio Grande.

Cortés gradually lost control of his empire, and in 1528 sailed for Spain in an attempt to have this authority reestablished by the King. Failing in this, he returned to New Spain in 1530 and began stocking, with cattle as well as sheep, the vast estates he had been granted in Oaxaca.

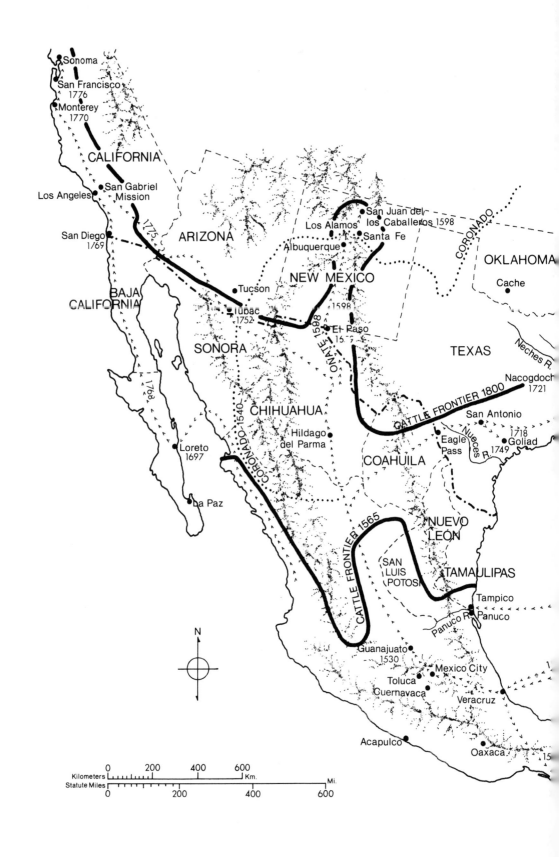

Sonoma
San Francisco
1776
Monterey
1770

CALIFORNIA

Los Angeles
San Gabriel
Mission

San Diego
1/69

1775

ARIZONA

BAJA
CALIFORNIA

Tucson

Tubac
1752

1768

SONORA

ONATE 1598

Los Alamos
San Juan del
los Caballeros 1598
Santa Fe

Albuquerque

NEW MEXICO

1598

El Paso

CORONADO

OKLAHOMA

Cache

TEXAS

Neches R.

Nacogdoch
1721

CORONADO 1540

CHIHUAHUA

Hildago
del Parma

Loreto
1697

La Paz

CATTLE FRONTIER 1800

COAHUILA

San Antonio

Eagle
Pass

Nueces R.

1718
Goliad
1749

CATTLE FRONTIER 1565

SAN
LUIS
POTOSI

NUEVO
LEON

TAMAULIPAS

Tampico
Panuco R. Panuco

Guanajuato
1530

Mexico City

Toluca
Cuernavaca

Veracruz

Acapulco

Oaxaca
15

N

0   200   400   600
Kilometers   Km.
Statute Miles
0   200   400   600   Mi.

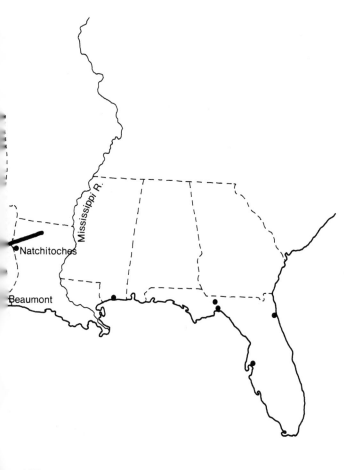

Natchitoches

Mississippi R.

Beaumont

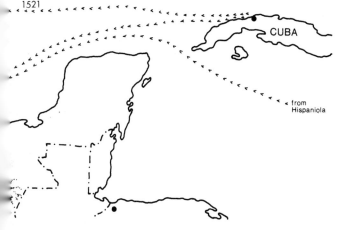

1521

CUBA

from
Hispaniola

Map 5.
Early Cattle Movement
Through New Spain

Again stock raising became the common vocation of retired military officers and the nobility from Spain who were
the recipients of large land grants. In the 1530's cattle ranches
were established as far north as the plains around Guanajuato.
(Brand, *Range Cattle Industry*) By 1539 wild cattle had reached
the present United States–Mexican border. In 1541, the well-
known expedition of Coronado reconnoitered the vast plains
region north of New Spain, from Arizona to Kansas. The commissary, along with other livestock, included a herd of cattle
numbered by some writers at 150 head, by others at 500. Coronado's animals, however, probably made no contribution to
breeding stock in the area as they would normally have been
consumed or have died along the route. The claims by some
writers that cattle left by Coronado established wild herds are
not authenticated.

The cattle industry flourished through the inhabited areas
of New Spain in the sixteenth century but grew phenomenally
on the high plateau. After the mountains surrounding Mexico
City are crossed, this region extends for a thousand miles to the
present United States border. The principal revenue came
from hides and tallow, the demand for which must have been
filled by only a small number of the natural increase. In 1586
an hidalgo named Ibarra was said to be running 130,000 head
and had an authenticated branding of 33,000 calves. Another
owner, Rio de la Loza, claimed a branding of 42,000. (Brand,
Range Cattle Industry) Why a grower would want to brand
such a multitude of calves when he could hope to sell the hides
of only a fraction of them is hard to understand.

There are many references to herds of 20,000 head and on
up to 40,000, even 50,000, on the central plateau of New Spain.
Juan de Oñate on his well-known expedition into the New
Mexico region in 1598 is credited with taking along 4,000 to
7,000 head. It should be noted, however, that such variations
could be due to an error in translating the word "*ganado.*" The
English equivalent is "livestock," not "cattle." Brand, however,
specifies "7,000 head of cattle (both ordinary and black cattle
. . .)" as being in Oñate's drive.

Ranching in New Spain progressed rapidly, always tending toward large operations. Open-range grazing, moving from one area to another as the grass growth necessitated, and semi-annual roundups for branding and selecting the offtake were the common practices. Bulls went uncastrated and breeding was uncontrolled. No more open field for the forces of natural selection to have full play can be imagined than that enjoyed by the herds on the central plateau of New Spain for the two centuries after their introduction.

Seldom is any description of cattle offered in old writings, and more than speculative reference to the fighting bull is even rarer. Of particular interest, therefore, is the report quoted below by one Spanish historian.

About 1552 Juan Gutierrez Altamirano, a cousin of Hernán Cortés, formed the Atenco Ranch in the Toluca Valley. Altamirano carried to that ranch 12 pairs of bulls and cows from Navarre with which he established the Atenco stock, reported to be the oldest fighting bull stock in Mexico. (Garcia Mercadal, Lo que España llevó a America)

Spanish bullfighting in those days was a sport of nobles on horseback. The bull was dispatched with a lance. The man on the ground with a sword had appeared, but this was considered to be a plebeian type of contest that attracted little attention. The true matador came in toward the end of the seventeenth century. The special rearing of bulls for fighting appears to go back to the time of the Moors in Spain. This mention of fighting cattle in New Spain in the sixteenth century is unique and indicates the exceptional nature of such a shipment.

References to "black" cattle by some narrators who touch on Spanish cattle in the early nineteenth century can usually be dismissed as a generic use of black, as "neat" in the term neat cattle. Black cattle are sometimes specifically noted, however, in historical accounts of activities in New Spain under such circumstances that it must be assumed that this was their color. Such references often indicate that these black animals were

55

less numerous than the common cattle whose color is not mentioned. This is quite conclusive evidence that herds of black cattle were segregated and maintained by some early ranchers in New Spain. It is not known when and how the nucleus of these herds was obtained. A newly arrived hidalgo in the sixteenth century from a black cattle area in Andalusia, who was in high favor at court and thus in position to have a special shipment authorized, is one possible hypothesis. Under favorable circumstances an importation of breeding stock could be arranged, as was done by an early viceroy for a cousin of Hernán Cortés in the movement of fighting stock mentioned above. Once established black cattle could have multiplied rapidly and been the means of founding herds whose owners were attracted by the novelty of the different color.

The Spanish Main

The northern coast of South America had been explored by the Spaniards since the opening years of the sixteenth century. Early attempts at founding settlements, however, were driven off by the Indians until the mid-1520's.

Colombia   A decade of exploratory effort and attempts at settlement had failed to establish a secured foothold along the northern coast of Colombia. While Santa María had been founded in 1510, it served mainly as a base for the colonization of Panama and was practically abandoned in 1517. After permanent settlements were established along the coast to the east, the town was rehabilitated in the 1530's and the Darién area appears to have been the source of the cattle that founded the first herds in the Sinú lowlands of northwestern Colombia.

Santa Marta, founded in 1525, and Cartagena, in 1533, were the ports from which the colonization of northern Colombia proceeded. Hispaniola was the seat of authority over all the north coast of South America at the time and was the point of origin for all support. The record is full of requests to the Crown from the coastal towns for cattle, permissions for importation, and orders by the *Casa de Contratación* to ship cattle. These attempts to stock the north coast continued from 1524 to 1537. A few cattle must have been received during this period as there are references to Indians slaughtering them. But complaints on the shortage, both for food and breeding purposes, continued.

Eventually cattle did arrive in sufficient numbers to become well established around the Santa Marta and Cartagena settlements and began to move inland. Thirty-five cows and a bull were shipped in 1543 from Santa Marta up the Magdalena River and then trailed through the mountains to the site of Bogotá. Two years later another consignment of cattle reached the same area. There were also at this time substantial movements of cattle from Lake Maracaibo in Venezuela to the Bogotá plains.

Cattle did not do well initially at the high altitude of Bogotá. In the 1550's laws were enacted prohibiting the movement of cattle out of the area, and restrictions were imposed on the kind of cattle that could be slaughtered. A period of fifty years appears to have sufficed for a fair degree of acclimatization. At the end of the sixteenth century the people on the high plateau were reported to have ample cattle for their needs.

Cattle reached the Colombian llanos by way of the Meta River from the Venezuelan llanos. The initial purchases that led to this movement were made around 1560.

Western Colombia was colonized through the back door from Ecuador. Sebastián de Belalcázar had joined Pizarro, and probably jealous of his commander's successes, wanted a gold discovery of his own. The Inca empire was broken and the mopping up operation had no appeal for the ambitious conquistador after his success in conquering Nicaragua. He took off on an unauthorized foray in 1534 to the unexplored area to the north. Traveling along the main cordillera of the Andes on

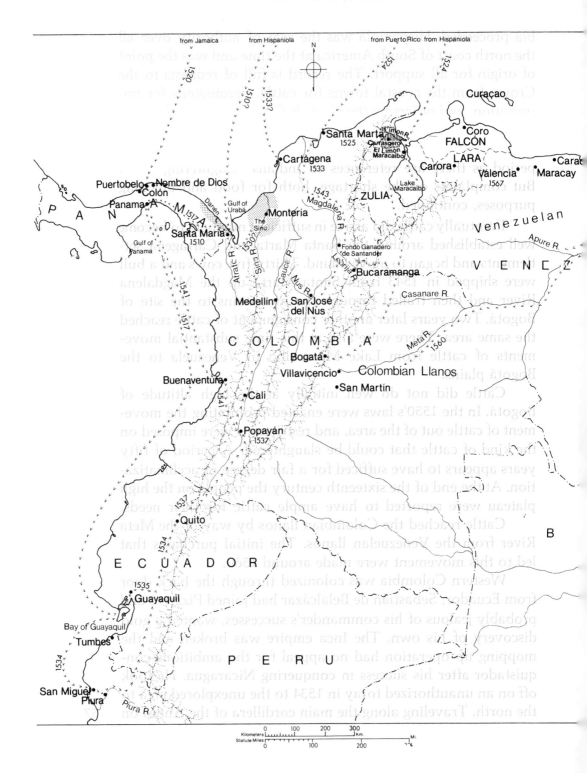

from Jamaica from Hispaniola from Puerto Rico from Hispaniola

Curaçao

Santa Marta
1525
Coro
FALCÓN

Cartagena
1533
Carora
LARA
Valencia
1567
Maracay
Carat

Puertobelo
Nombre de Dios
Colón
Panama
Lake
Maracaibo
ZULIA

Santa Maria
1510
Monteria
The
Sinú
Venezuelan
Apure R.

Fondo Ganadero
de Santander
Bucaramanga
VENEZ

Medellin
San José
del Nus
Casanare R.

COLOMBIA
Meta R.
1560

Bogatá
Villavicencio
Colombian Llanos

Buenaventura
San Martin

Cali

Popayán
1537

B

Quito

ECUADOR

Guayaquil
1535

Bay of Guayaquil
Tumbes

PERU

San Miguel
Piura
Piura R.

0   100   200   300
Kilometers   Km.
Statute Miles
0   100   200   Mi.

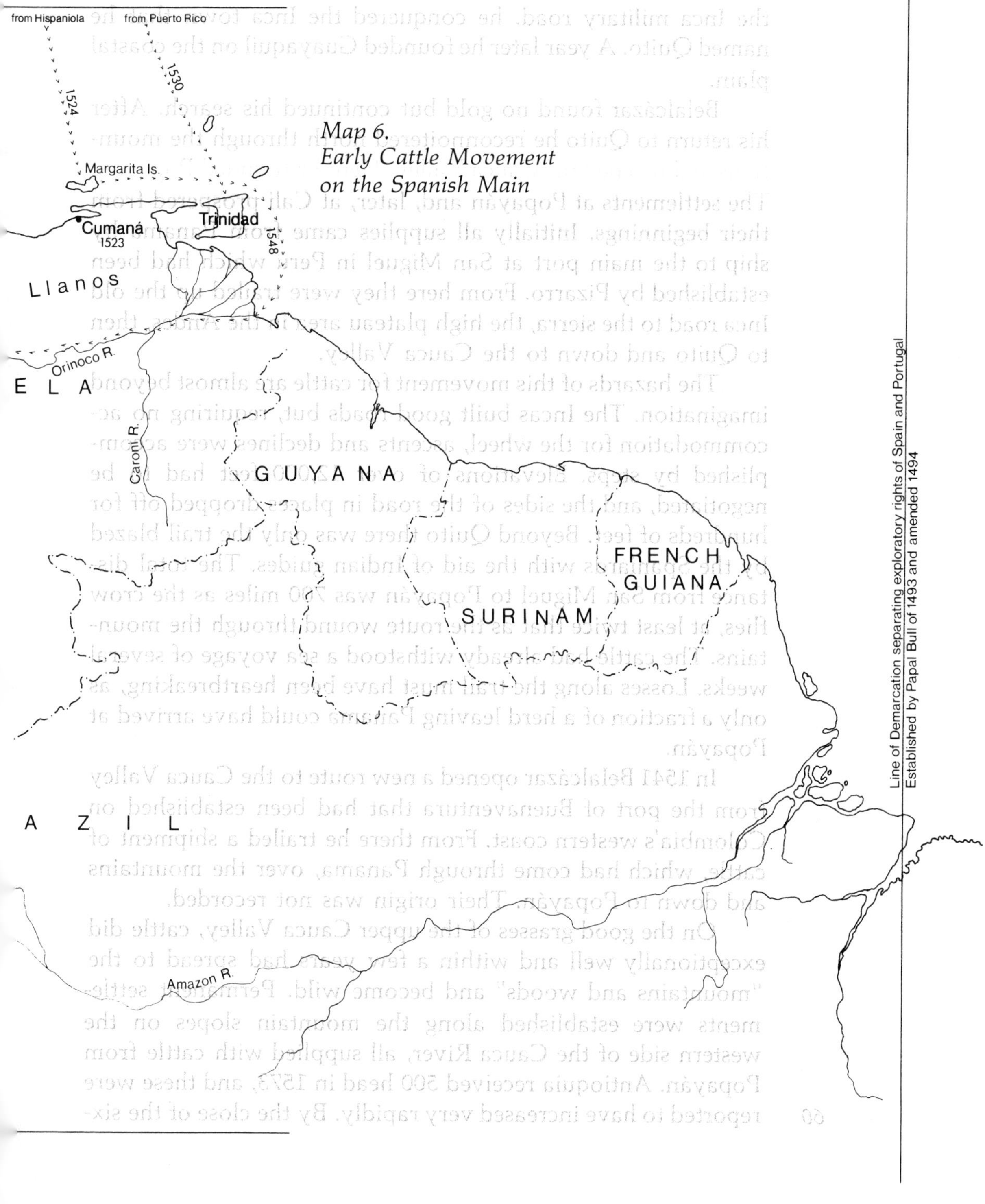
from Hispaniola
from Puerto Rico
1524
1530
Margarita Is.
Cumaná
1523
Trinidad
1548
Llanos
Orinoco R.
E L A
Caroni R.
GUYANA
FRENCH GUIANA
SURINAM
A Z I L
Amazon R.
Line of Demarcation separating exploratory rights of Spain and Portugal
Established by Papal Bull of 1493 and amended 1494

Map 6.
Early Cattle Movement on the Spanish Main

the Inca military road, he conquered the Inca town that he named Quito. A year later he founded Guayaquil on the coastal plain.

Belalcázar found no gold but continued his search. After his return to Quito he reconnoitered north through the mountains and reached the Cauca Valley. Here he founded Popayán. The settlements at Popayán and, later, at Cali prospered from their beginnings. Initially all supplies came from Panama by ship to the main port at San Miguel in Peru which had been established by Pizarro. From here they were trailed up the old Inca road to the sierra, the high plateau area in the Andes, then to Quito and down to the Cauca Valley.

The hazards of this movement for cattle are almost beyond imagination. The Incas built good roads but, requiring no accommodation for the wheel, ascents and declines were accomplished by steps. Elevations of over 12,000 feet had to be negotiated, and the sides of the road in places dropped off for hundreds of feet. Beyond Quito there was only the trail blazed by the Spaniards with the aid of Indian guides. The total distance from San Miguel to Popayán was 700 miles as the crow flies, at least twice that as the route wound through the mountains. The cattle had already withstood a sea voyage of several weeks. Losses along the trail must have been heartbreaking, as only a fraction of a herd leaving Panama could have arrived at Popayán.

In 1541 Belalcázar opened a new route to the Cauca Valley from the port of Buenaventura that had been established on Colombia's western coast. From there he trailed a shipment of cattle, which had come through Panama, over the mountains and down to Popayán. Their origin was not recorded.

On the good grasses of the upper Cauca Valley, cattle did exceptionally well and within a few years had spread to the "mountains and woods" and become wild. Permanent settlements were established along the mountain slopes on the western side of the Cauca River, all supplied with cattle from Popayán. Antioquia received 500 head in 1573, and these were reported to have increased very rapidly. By the close of the sixteenth century the Sinú, the llanos and adjoining slopes of the eastern cordillera, the Bogotá plateau, and the Cauca Valley— all the principal cattle raising areas of Colombia—were well stocked.

The introduction of cattle to Venezuela followed much the same pattern as in northern Colombia—a surfeit of requests, permissions, and orders and a paucity of recorded landings. Cumaná was the first permanent settlement, established about 1520, and, while there appears to be no record of cattle being landed there in the early days, in 1524 a shipment from Hispaniola arrived at Margarita Island, just off the coast. (Patiño, *Plantas y Animales in América*) Cattle from Hispaniola had reached the Lake Maracaibo area by 1534. A trail was blazed for the movement of cattle to Bogotá, and later an active commerce in cattle between Venezuela and Colombia developed.

Venezuela

Most of the nearby islands lying off the northern coast of Venezuela were settled by Spaniards early in the second quarter of the sixteenth century. Cattle were introduced, and the largest island, Curaçao, became a recognized center for cattle trading in those days. When the Dutch became active in the Netherlands Antilles in 1635, the islands were already well stocked with Spanish cattle.

The first cattle to reach the llanos of eastern Venezuela were descendants of those that had been on Margarita Island for more than twenty years. They arrived at the mouth of the Orinoco River about 1548. (Humboldt, *Equinoctial Regions*) Settlement began at the confluence of the Orinoco and Caroni rivers and then proceeded westward along the Orinoco to its junction with the Apure. By 1560 the herds had reached the Meta River at the border that was eventually established with Colombia.

Cattle raising was the sole occupation of llanos pioneers. The animals became adapted to the inhospitable environment, but how long a period of adjustment was required has not been recorded. For generations it was a matter of survival of the fittest. The llanos had the poorest forage that Spanish cattle had grazed since reaching the New World. The savannahs were flooded in the rainy season, and, although the dry grass was more nutritious after the rains ceased, the water holes dried up and forage became scarce where cattle ran. But the herds eventually increased and then multiplied rapidly. The Andalusian methods of management had come with the cattle but could do little to provide better pasture when needed. Natural selection took over for three centuries and produced the Llanero, the Criollo of the llanos, as hardy a bovine as ever lived.

Because of trouble in subduing the Indians, Valencia was the last settlement established in the days of Venezuelan colonization. After the Indians were finally overcome around 1567, the area proved to be good cattle country. In all the other settlements in Venezuela, cattle were prospering by the last half of the sixteenth century. There were no untoward happenings to hinder the growth of the cattle industry in the country until the start of the wave of independence from Spain after the opening of the nineteenth century.

One additional, but minor, introduction of cattle to Venezuela should be noted, as it is one of only three known instances where African cattle reached American shores. In 1536, during the brief period of German occupation of western Venezuela, a Portuguese colonizing party landed some cattle which had been picked up in Africa. (Patiño, *Plantas y Animales en América*) The endeavor failed, and it is doubtful if any cattle survived.

Trinidad   As a defensive measure Spain made every effort to hold the islands that abutted the coast of South America. This was in contrast to her lack of interest in the small islands of the Lesser Antilles. Trinidad, as has been seen, was discovered by Columbus on his third voyage. It was designated by royal order as one of the islands on which the Carib Indians could be taken as slaves.

A few cattle were received by the early settlers in 1530, the shipment coming from Puerto Rico. After a governor was appointed in 1532, shortages of cows were repeatedly reported to the Crown. Requests for more cattle were made to the *Casa de Contratación* during the next decade, but few were received. Finally, in 1599 the governor reported that there were "enough cattle for the people and also to supply some to the Guianas." (Patiño, *Plantas y Animales en América*)

There were only Spanish cattle on the island when the British gained possession at the end of the eighteenth century. As the Spaniards had an exportable surplus there 200 years earlier, there was no occasion for the British to add to the herds existing at the time.

Off the true course of the Spanish Main, the northeast coast of South America was largely ignored by Spain. Permanent colonization of the Guianas began at the end of the sixteenth century. Holland, England, and France then began competing for footholds in the only part of South America that was still open to conquest. The area of the Guianas was lacking in productive grasslands, and when agricultural development got underway, cattle were in demand mainly as draft animals in the sugar operations.

The first cattle from which breeding herds were established were brought to the Guianas late in the sixteenth century from Trinidad. These probably made their contribution to the foundation stock of Spanish cattle that eventually populated the region. When the British in the mid-seventeenth century started to colonize the area that is now Surinam from Barbados, cattle were brought in from that island. These animals would also have been of pure Spanish descent.

Cattle from Brazil, which were descendants of the original Portuguese stock that founded the herds of that country, did reach the Rupununi district of western Guyana. The Repununi is well isolated by rain forest from the agricultural coastal areas of the Guianas, and it is doubtful if the Portuguese cattle contributed materially to the genetic makeup of the Guiana cattle. Crossing the Criollo herds with Zebu bulls did not begin until after World War II.

Guianas

63

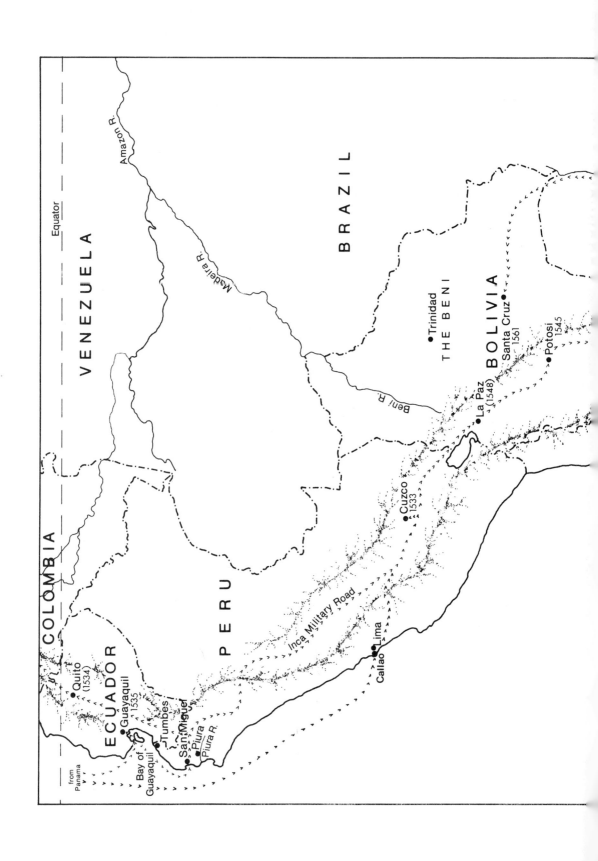

COLOMBIA

VENEZUELA

Equator

Amazon R.

Madeira R.

BRAZIL

Beni R.

Trinidad
THE BENI

BOLIVIA

La Paz
(1548)

Santa Cruz
1561

Potosí
1545

Cuzco
1533

Inca Military Road

Lima
Callao

PERU

Quito
(1534)

ECUADOR

Guayaquil
1535

Tumbes

San Miguel

Piura
Piura R.

Bay of
Guayaquil

from
Panama

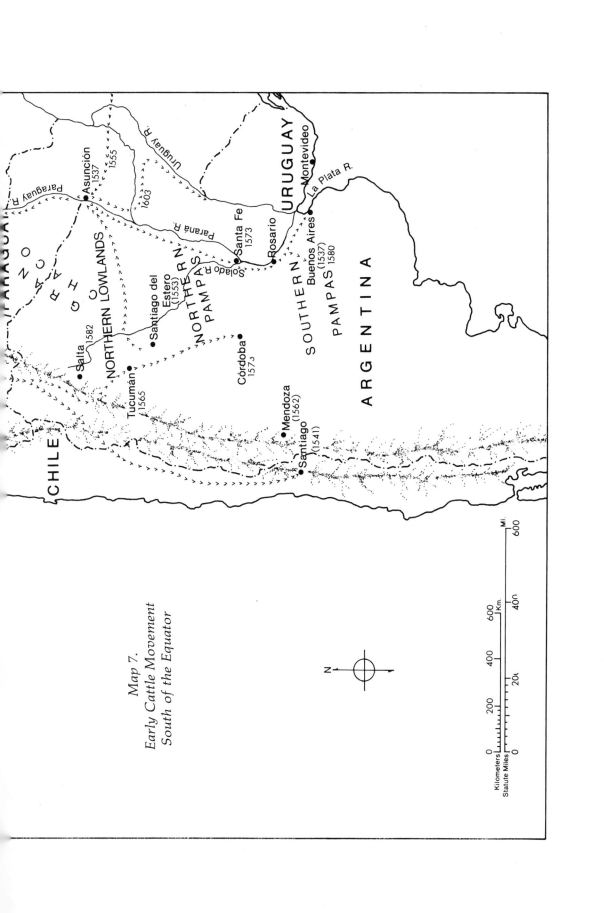

Map 7.
Early Cattle Movement
South of the Equator

The Inca Empire and Beyond

A boyhood swineherd in southwestern Spain, Francisco Pizarro, had worked his way to the Indies shortly after their discovery and became a lieutenant of Pedrarias in Panama. With the aid of another conquistador and a friar, Pizarro managed to organize a small exploratory expedition that sailed down the west coast of South America in 1524 and stumbled onto traces of the Inca empire. This led to the conquest of Peru in which Pizarro unwittingly blazed the trail for the cattle industry that was to extend from the Cauca Valley of Colombia to the pampas of the Argentine.

That a swineherd paved the way to the cattle empires of the Argentine does not appear to have been noted by historians. Here was one of the earliest rags-to-riches stories of the New World, although Pizarro was killed before he could enjoy the fruits of his successes. Most Spaniards who rose to fame in the new lands were men of family and means.

Peru,
Eduador,
Chile

We have seen how, in a diversionary movement of Pizarro's conquest of the Incas, Belalcázar had conquered Ecuador and moved on to establish the Spanish presence in the Cauca valley of Colombia. After Guayaquil had been established, cattle were shipped in from Panama to found breeding herds. A number of haciendas with sizable cattle holdings came to occupy the surrounding coastal plains. Isolated by the encircling mountains, ranching operations were of little economic importance until modern times because of lack of a market. Around Quito the early farmers had their draft ox and milk cow, but cattle must have required many decades to become acclimated in the high altitude. Ecuador never became a cattle country, and holdings did not expand beyond what was required locally for food and draft power.

The backbone of the Inca empire was broken when Pizarro took Cuzco in 1533. The Spanish conquerors then proceeded to fight among themselves over the spoils for another decade. In

66

spite of these fratricidal conflicts colonization followed in the wake of the conquistadors.

Both the conquest and the colonization were made possible by the old Inca highway. It provided access to areas that otherwise would have been inaccessible. From the port at San Miguel the conquistadors had made an easy trail to the foot of the western slope of the Andes where it intersected the highway. From this junction the route led south for 2,000 miles along the cordillera and then down to the Argentine lowlands. The road north of the junction was that taken by Belalcázar when he moved on Quito. These were actually paved highways built by Inca engineers—military routes that led to all parts of their empire. The Inca warrior was a foot soldier, and the llama was the standard pack animal. The paved roads were wide enough for the passage of several animals abreast through level valleys, but when winding around a high cliff or through a narrow defile cut through solid rock, footing was provided for only one man or beast. Steps were employed for steep grades. Fast mountain streams were crossed by bridges built of vines hanging across canyons hundreds of feet deep. The fiber tread was sufficient for the light-footed llama but was soon cut to shreds by the hoof of horse or ox. The Inca bridges which had served the Spanish military for crossings were useless when the supply trains began moving to the mines. Here a ford had to be found or a raft built. Access to the stream required construction of a new trail down to the water and up again on the other side. Blind turns, where all an animal could see ahead was a canyon bottom a few thousand feet below, must have been a problem in trailing cattle.

Driven by their continual lust for gold, the Spaniards traveled along the Inca military road to its end high in the mountains above Chile. From here, with the aid of Indian guides, a trail was finally made through the precipitous gorges of the Andes down to the coastal lowlands. Many Spaniards, as well as their Indian attendants, died in the high altitudes because of storms and shortage of food. The site of Santiago was reached in 1536, but permanent habitations were not built until five years later.

A few cattle reached Chile in the early days of settlement. Pedro de Valdivia, the leader of the successful colonizing expedition, in his report to the King, mentioned himself as ". . . the head shepherd in the breeding of cattle." (Winsor, *Spanish Exploration and Settlement*) The area, however, was not attractive to stockmen until sheep raising was initiated in the far south. The struggling settlements in the low valleys on the Pacific side served as a base for the expeditions that founded the first settlements in northern Argentina.

After Santiago was established, the conquistadors retraced their steps across the mountains and worked their way down to the northern Argentine pampas. Santiago del Estero, the first permanent abode of the European in the Argentine, was founded in 1553.

In the meantime fabulous silver lodes had been uncovered at Potosí in the high mountains around the old workings where the Incas had mined their silver. A wild stampede started to the altiplano in 1545. Former conquistadors, miners, and disillusioned adventurers from the established colonies in the Indies—all left for the mines. The route now was by ship from Panama to the west coast port of Callao, which had been founded in 1537, and then through the mountains to Potosí. When La Paz was established in 1549, the rush to the mines increased.

The Andalusian cattleman had no experience in raising cattle in high mountain areas, and breeding herds were not established around Potosí. The demand for beef in the fast-growing mining settlements could not be met by the gaunt cattle which had struggled over the 1,400-mile trail from the coast, where they had been landed after a two-month voyage from Panama.

Mules, always a favored pack animal of the Spaniard, had been found far superior to horses for mountain transport and were in even greater demand than cattle. The faithful ox could not pull his cart up the steps of the Inca highway nor could a wheeled vehicle negotiate the narrow defiles along the cliffs. For twenty years the mines struggled along, dependent for livestock on the thin supply line from the coast. Finally it was realized that the northern lowlands of the Argentine were the logical supply point for livestock to support the mines.

Eventually settlers in the altiplano of Peru and Bolivia acquired small herds which became acclimated to the high altitude. Some ranching operations were established south of San Miguel in the Piura River drainage. The central valley of Chile, where Santiago became the focal point, was a hospitable home for cattle but operations were limited to farm herds. The west coast of South America did not have the wide grasslands to become a real cattle country.

The first stockraising operations in the Argentine began in the northern lowlands. This is a region of savannah and scrub forest with wide flood plains in the rainy season—not an ideal range for cattle but one to which the Spanish cow soon adapted.   **Argentina**

Tucumán, at the edge of the lowlands, was conquered by the conquistadors in 1550, but the Spanish settlement was not founded there until 1565. The supply line to the Andean mines was by ship from Panama to the port of Callao, then along the old Inca road to Potosí. When Tucumán became a livestock center, the trail for mules and cattle to the mines was shortened to 700 miles as against twice that distance to the mines from Callao plus a two-month sea voyage from Panama. The advantages of the northern lowlands as the source of livestock and related products for the Potosí area are obvious but they went unutilized for years.

The demand of the mines for mules and beef eventually led to stocking the Tucumán area. Here the first breeding herds in Argentina were established. It was nearly two centuries later when the scene shifted south to the pampas and wild cattle were hunted and killed for their hides.

For years the most urgent need was for mules to pack in supplies from the coast and to work in the mines. The foundation herds were built from mares and asses that had to be moved in from Panama. As the demand for mules was filled, cattle were brought in, and breeding herds established to supply the mines with beef. Progress in cattle raising was slow initially. Losses in moving the breeding stock were disheartening. The voyage down the coast from Panama, then by land across the mountains, required upwards of six months under severe conditions. The survivors, once in the grassy lowlands, did well, however, and soon increased to the point where they were running wild, following the usual pattern of the Spanish cattle wherever they were introduced in the New World. Salta later became the main gathering point and market for cattle destined for the mining districts.

For two centuries, the northern lowlands were the source of the essentials of life to the settlements in the Andes. Grain, mules, and the human needs derived from cattle—live animals for beef and hides for innumerable uses—all went up the trail to the mines.

The manner in which cattle reached the pampas is not entirely clear. An attempt in 1536 to plant a settlement at the site of Buenos Aires had been abandoned after a five-year attempt to hold off hostile Indians. The survivors then moved inland to Asunción. Horses were left behind from which the wild herds on the pampas became established, where they were found years later. No mention is made of cattle being included in the stock that was abandoned, and it must be concluded that any which the expedition had initially were consumed. Although Buenos Aires was reestablished in 1580, it was only a military outpost until made a viceroyalty two hundred years later. During this interval, cattle operations were confined almost entirely to the northern lowlands. There was some contraband trade in hides out of Buenos Aires but this did not reach sizable proportions until a few years before the viceroyalty was established.

The first cattle of record on the southern pampas came from Asunción. In 1580, when an expedition left to reestablish the settlement at Buenos Aires, 500 head of cattle and 1,000 horses were included. The cattle were both for food en route and to establish a breeding herd when the destination was reached. Poor trail management resulted in a large number of both horses and cattle straying, the cattle to found the original wild herds of the pampas, and the horses to add to those that were already there. Many generations later, their wild progeny were hunted for their hides by nomadic gauchos.

When Buenos Aires became a viceroyalty in 1776, large numbers of cattle covered the pampas. Range-type operations were founded on the large land rights granted to retired conquistadors and hidalgos from Spain. These land rights also included authorization to appropriate wild cattle and horses. Ranching progressed rapidly as Buenos Aires became a major port and an outlet for the export of hides, tallow, and dried beef. The beginning of the nineteenth century saw the great grassland of the Argentine pampas well stocked with cattle, the origin of which had been in the Indies.

Bolivia

Except for the altiplano mining district along the western border with Peru, Bolivia was well off the course of Spanish conquest. Most of the country, lying in the lowlands of the upper Amazon basin, known as the oriente, was not a productive source of gold and was given little attention. Santa Cruz de la Sierra, the first colony east of the Andes, was founded in 1561 by an expedition that had moved up from Asunción. Cattle to found the first breeding herds were probably trailed in from northern Argentina.

The oriente has remained one of the most isolated regions of South America down to modern times. Sparsely populated from the time of the first settlements, lacking minerals, and with no outlet for agricultural products, little attention was given to stock raising. The Spanish cattle which accompanied the early settlers supplied their immediate needs and soon came to exist in a wild state. Through natural selection they became ideally adapted to the hot, humid climate and remained uncontaminated with other cattle types until well into the twentieth century.

Paraguay Although the Asunción settlement had been founded in 1536 by an expedition which moved up the Paraná River from Buenos Aires, the first cattle did not arrive in Paraguay until 1555. At this time seven cows and one bull were trailed in from the east coast of Brazil. (Warren, *Paraguay*) These were probably descendants of cattle which had originated in Portugal. A shipload of cattle is reported to have moved up the Paraná River to Asunción in 1569 which would have been of Spanish origin. To what extent the herds of the well-established ranching operations around Tucumán and Córdoba contributed to stocking the Chaco of Paraguay does not appear to have been recorded. In colonial times Asunción was off the main line of commerce which lay between the mines and the northern lowlands. There was little demand for cattle products, even for hides or tallow.

Early in the seventeenth century, a line of Jesuit missions in southern Paraguay was built and prospered for 150 years. Considerable agricultural progress was made, and large herds of cattle were developed and cared for under controlled range conditions. Indians were trained to become herdsmen. The missions were abandoned in 1767, and the livestock fell into the hands of inexperienced settlers. Beyond filling the local needs for beef and hides, the cattle industry in Paraguay made little progress until the nineteenth century.

Uruguay In 1603, the governor of Paraguay made a sizable shipment of cattle and horses—some reports say 100 head of each—down the Paraguay River to be unloaded on the Uruguay side. This may have been a farsighted move to establish a source of livestock for a future operation in the coming feud over boundaries with Brazil. The stock ran free on the virgin savannahs and the increase in cattle, as well as horses, was fantastic.

A century later the roaming gauchos from the Argentine began to cross the Uruguay River and kill cattle for their hides. Here animals were more plentiful and easier to harvest than those on the west side of the river. Thus began the Uruguayan cattle industry. This practice led later to the sale of cattle for

72

slaughter to traders from Argentina. Until after the opening of the nineteenth century, this trade based on wild cattle was the only livestock activity of a commercial aspect. Montevideo, founded in 1726, was only a garrison town until the end of the century, established to hold off Brazilians.

The Uruguayan grasslands, the most productive in all South America, were the last large rangeland in South America to see the development of an organized livestock industry. The area had a surplus of both wild and domestic Spanish cattle all through the eighteenth century.

Florida, Louisiana, and Lands North of the Rio Grande

Ponce de León, seeking wider fields after his success as governor of Puerto Rico, obtained royal permission to explore Florida. The region at that time was thought to be an island. He out-fitted an expedition at his own expense and made a landing in 1513 on the east coast near the future site of St. Augustine. The party was soon driven off by Indians of a fiercer type than those that had been encountered on the islands. Although the exploration was continued down the coast and around to the west side of the peninsula up to Tampa Bay, no other landings were reported.

De León made a second attempt at Florida eight years later. Apparently he had only two vessels, fitted out for colonization. It is evident that cattle were carried, the number given by different writers varying from 50 to 200 head, even the lesser of which is ridiculously high for only two ships equipped to found a settlement. The prospective settlers were again driven off by the Indians and undoubtedly whatever cattle they had were left behind. These were probably the first bovines on Florida soil, although some writers record that cattle were carried on de León's first voyage.

Five other attempts over the next forty years, some by

Florida

73

well-outfitted expeditions, likewise failed to gain a foothold in Florida. The Indian warriors, good fighters and well equipped, prevented any permanent settlement. In the records of these later expeditions, mention is sometimes made of cattle being included in the cargo, but these had to be abandoned when the conquistadors departed. It would indeed be surprising if some of these attempts at settlement did not leave small groups of breeding cattle behind. Given the Spanish cow's propensity for survival, and the certainty that when an expedition failed any remaining cattle would be abandoned, the possibility of a small herd of wild cattle establishing themselves on Florida soil cannot be denied.

All of these early expeditions, however, were driven off by Indians. And the Indians loved beef. A few head of stray cattle could easily have ended in a celebration barbecue. It must be remembered that at this time the Indian had not learned from the white man, as he later did, the art of tending cattle. Such considerations are a valid counter to the wild-cattle hypothesis. Whether wild cattle existed in Florida when Pedro Menéndez de Áviles founded St. Augustine will remain an open question until further evidence comes to light.

A French invasion on the east coast of Florida in 1562 aroused the Spaniards to resolute action. The King gave Menéndez three ships and 600 military to drive the French off and a contract to colonize and govern Florida. Menéndez, an old pirate hunter who had secured the sea lanes from Spain to the Canaries, furnished an additional eight vessels and 900 men on his own. Leaving from Cuba, his force landed in Florida in 1565, killed the French garrison, and founded St. Augustine, which has long been recognized as the first permanent settlement of Europeans on what is now United States territory.

The St. Augustine colonists brought with them what were probably the first breeding cattle to become established in North America. These were Spanish cattle from Cuba. Outlying missions were founded and attempts made to Christianize the Indians, attempts that were only partly successful. The colony hung on but did not enjoy any great prosperity.

The second settlement area in Florida was established by 1655 near Tallahassee and was soon stocked with Spanish cattle. Several outlying missions were built and settlers took up farming. Here cattle did exceptionally well. Florida was still primarily a military outpost, but the garrisons increased in size, and farmers had a substantial income from beef sold to the forts.

Spanish cattle undoubtedly reached as far north as Chesapeake Bay during the seventeenth century. After the St. Augustine and Tallahassee areas had been secured, the Spaniards did their best to establish a defensive frontier north of Florida. These efforts were continued throughout most of the century. A number of small missions were scattered through Georgia, Alabama, and up into the Carolinas, and there was at least one mission in Virginia. These early outposts of the Church certainly included a few head of cattle, as they were always an essential adjunct to the missions. The English military and pioneers, probing into the same area, often aided by the Indians, prevented any permanent Spanish settlement. Although missions would be built, within a few years the residents were driven out or the projects abandoned because of the difficulty of supply, and by the end of the century few, if any, traces were left of the attempted Spanish occupancy.

The presence of the Spaniard at this time in the area of the southern English colonies along the Atlantic seaboard is the basis for the assumption by some writers that Spanish cattle in a wild state existed in the region. The missions, however, were short-lived and never prosperous. Their cattle holdings must have been small to start with, and none of them existed long enough for sizable herds to be built. The peak activity of the cowpen people of the Carolina piedmont was reached in the mid-eighteenth century — many decades after the last Spanish mission north of Florida had been abandoned. It is doubtful if any of the small cattle holdings of the padres survived the predators and Indians long enough to fall into the hands of the Carolina cowboys.

Early in 1698, aware that France was planning to embark

75

on the gulf coast west of Florida, Spain moved to strengthen her flank in that area. Pensacola was established by expeditions sent out from both Havana and Veracruz. Cattle for the commissary and for founding a breeding herd probably came from both Cuba and Mexico. This was a more favorable environment for stock raising than the parts of Florida previously settled, and the foundation herd increased quite satisfactorily. When Iberville landed several months later off Mobile Bay, the French cast envious eyes on the Spanish herd at Pensacola. The opening of the eighteenth century saw farming communities well established around St. Augustine and Tallahassee, with individual herds in the 50- to 100-head range, as estimated from their owners' tax returns.

After the English harassment of Florida began in 1702, the pillage in the wake of war and Indian uprisings completely disrupted the economy. Breeding herds were killed off and not replaced. This period of stagnation existed until Florida was surrendered to England in 1763. The record is far from clear as to what happened to the Florida cattle industry during the two decades of English occupancy. Under the more settled conditions and with an increase in population, cattle numbers must have expanded. Some of the Indian tribes had now learned the rudiments of cattle raising from the white man and had managed to obtain herds of their own.

When Florida was ceded back to Spain in 1783, there was a sizable efflux of recent settlers. The loss of cattle in this movement was more than offset by the Spanish efforts to encourage settlement. Land grants, many of large acreages, were freely given to substantial colonists to encourage immigration. Cattle raising in many parts of Florida became the most important sector of agriculture, because it provided marketable products in areas where transportation was limited to the oxcart or small river boats. Hides, tallow, and salted beef could always be got to market. Another factor was that clearing and burning could very quickly bring large areas of the terrain into production as pasture. Cultivation for crops took more time and effort. In 1821, when Florida became United States territory, it was stocked with substantial numbers of pure descendants of the Spanish cattle—the Florida Scrub, sometimes called Florida "cracker."

Spain had made no move to claim the area around the Missis-Louisiana
sippi River delta and the country to the north, although a Spaniard, de Soto, had discovered the river in 1541. The French began colonization around Mobile Bay at the end of the seventeenth century, which led to the founding of New Orleans.

Pierre Lemoyne, Sieur d'Iberville, a native French Canadian with a successful record in fighting off the British around Quebec, in 1698 obtained a patent from the French King to colonize the Mississippi delta area. Sailing from France a year later, Iberville and his brother landed 200 settlers on Dauphin Island off the coast of Mississippi. These colonists later moved to Ship Island and then to the mainland near present-day Biloxi. The Iberville party had a few cows and bulls, along with other livestock, all of which came from France. The cattle probably were consumed long before they could produce replacements.

The nearby Spanish colony at Pensacola had cattle at this time, but none of these got into the hands of the struggling French settlement. The French were well aware of the success of the Spaniards with cattle throughout the Indies and made every effort to obtain foundation stock from Cuba and Hispaniola. Additional shipments from France apparently were not considered feasible because of the distance and the danger from privateers.

The Spaniards were naturally reluctant to help the rising French community that had sprung up in the midst of their empire. In 1701 Iberville managed to buy a few head of cattle in Hispaniola for his settlement. But in 1703 a French ship sent to Havana came back with only four oxen, the only cattle the Spaniards would sell. Another occurrence a few years later emphasized the Spanish attitude. Sixty cows had been secretly purchased around Havana by Frenchmen. When they were loaded on the ship chartered to move them, the Spanish   77 authorities interfered and forced all but fifteen head to be unloaded before the vessel was cleared.

Efforts to get cattle from New Spain, Texas, and Florida, as well as from Cuba and Hispaniola, met with only small success. Approaches to the English colonies for cattle failed even though at this time there was a large population of Native American cattle in the Carolinas and Georgia. The persistence of the French finally succeeded in establishing the foundation of a breeding herd. "By 1708 the stock of the colony consisted of 50 cows, 40 calves, 4 bulls and 8 oxen." (Gray, *History of Agriculture*) These had increased to 400 head by 1717, a reasonable increase from fifty cows, but still not many cattle for the rapidly growing colony.

For years after New Orleans was founded in 1718, the colony was dependent on meat, largely dried buffalo, brought down the Mississippi. As settlers began moving onto the grasslands of southwestern Louisiana, the area became well stocked with Spanish cattle. Brand registration began in 1739. In the period from 1760 to 1888, there were 27,000 brands recorded for the southwest district of Louisiana alone. Near open-range conditions prevailed except for the small herds owned by farmers. The cattle industry flourished through the last quarter of the eighteenth century on the range lands which reached from the Mississippi west to Spanish Texas.

There was one known increment of northern European cattle brought to Louisiana before the influx of Native Americans began. When the Acadians, expelled from Canada in 1755, eventually reached Louisiana, they brought with them their French Canadian cattle. The progenitors of those bovines had been raised in Canada for 150 years. For a time these were carefully maintained in Louisiana in the farmer's barnyard as dairy animals, but eventually they became absorbed in the large population of Spanish cattle and left no discernible trace.

When Spain gained possession of Louisiana in 1762, there was no change in the cattle industry as it had developed on the open prairies. The French vachers, who had learned range management of cattle from their meager contacts with the Spanish

78

vaqueros, had rapidly expanded their herds. The Spanish authorities were well content to let the French settlers continue in their agricultural pursuits. Herds in the 1,000- to 2,000-head bracket were common, some even up to 4,000 head. Boats working their way down the Mississippi in the New Orleans trade carried loads of 300 to 400 head from the cattle markets upstream.

At the opening of the nineteenth century, the herds in Louisiana were pure Criollo. Locally they were still called Spanish cattle. Up to this time there had been practically no contact with the Native American cattle to the north and east.

From 1540 to 1542 Coronado had traveled out of New Spain through New Mexico and the western plains on his renowned exploration. The assumptions that his commissary herd of 500 head established wild cattle along the route are pure fantasy. The head count of Coronado's expedition has many discrepancies in the record but can be "rounded off" at 1,000 Indians, 100 foot soldiers, 500 cattle, and 250 men on horse. The cattle, along with a large number of sheep and goats, were brought along for food.

New Mexico

It seems as though 250 Spanish horsemen could have kept 500 head of cattle from straying. The well-known ability of the mounted Spaniard to handle cattle should have prevailed even if the horsemen were cavalrymen and not vaqueros. Assuming that a few head did get away, they would soon have fallen prey to the Indians, and therefore, the ultimate survival of any of Coronado's cattle cannot be credited. During the fifty years after Coronado's foray, other exploratory expeditions followed his trail in search of the legendary Seven Cities of Gold. No presidios or missions were founded, and the efforts were generally conceded to be failures. No cattle inadvertently left behind would have survived to propagate.

The first authenticated breeding herd north of the Rio Grande was the Spanish cattle that Juan de Oñate trailed to northern New Mexico in 1598. Oñate, a native, and one of the wealthiest men in New Spain because of the family silver mines, had been appointed governor. He was authorized to colonize the region at his own expense. Oñate established his frontier capital at San Juan de los Caballeros, the first permanent settlement north of the Rio Grande. His party had traveled up from southern Chihuahua to the river, followed it to El Paso then north almost to the present Colorado line.

This expedition included a cattle herd, variously placed at 4,000 to 7,000 head by different writers. Even 4,000 appears high; 2,500 head were about the maximum that experienced drovers took on the long drives from Texas to Kansas in the 1880's. There can be no question, however, that Oñate did take a large number of Spanish cattle to northern New Mexico and that these founded many breeding herds.

The capital suffered continued harassment from the Indians and about 1609 was moved to Santa Fé. Missions were established in the pueblos of the Indians who were instructed in the rudiments of agriculture and stock raising as practiced by the Spaniards. For three quarters of a century the padres had reason to feel that their efforts to bring the Catholic faith to the heathen had been successful.

The missions were highly successful with their cattle, and their herds ran into the thousands. Large haciendas with sizable cattle holdings were established under private ownership. The Indian population of the settlement, however, far outnumbered the Spaniards. A decade after the move to Santa Fé, 30,000 converts were reported under the dominance of only a few hundred soldiers, priests, and settlers. Colonization proceeded under the established pattern of demanding food and labor from the inhabitants of conquered lands. In 1680 the subdued antagonism of the Indians finally erupted in a general rebellion. Caught unprepared, the Spaniards were driven back to El Paso del Norte. The ranches with their large herds were abandoned. The padres, however, had taught the Indians the husbandry of cattle, and in one way or another, many of the thousands of head in the area survived until the reentry of the Spaniards

sixteen years later.

A punitive expedition resecured Santa Fé in 1696. This was the beginning of permanent colonization in New Mexico. Stock raising, sheep and cattle, was the base on which progress was made and expanded throughout the region for the next hundred years. The semiarid grasslands provided excellent pasture, although large areas were necessary for successful ranching. The settlers were scattered over a wide territory, and the few towns that grew up were spread over much of the region. This expansion spread into southern Arizona, which was part of the Spanish province at the time.

By the opening of the nineteenth century, New Mexico was better stabilized than any of the Spanish colonies in what is now United States territory. The population was estimated at 20,000, not including the Indians. These colonists were nearly all descendants of the conquistadors, hidalgos, and other settlers who had moved up from New Spain in the two centuries that New Mexico had been under development. The similarity between the wide grazing lands and moderate climate of New Mexico and the highlands of Castile, where many of the original conquerors had hailed from, was undoubtedly a factor in the attraction New Mexico had for the Spaniards.

Eventually New Mexico proved to be better sheep than cattle country. The Castilian grandee reverted to his old love of sheep. Cattle were relegated to a minor role in the economy of this frontier. When the western range lands were being stocked with the Longhorn in the latter half of the century, it was necessary to bring breeding stock back into New Mexico.

Arizona

The more arid lands of Arizona (part of Spanish New Mexico) were not as attractive to the Spaniards as the better range to the east. Conquistadors and the fathers in the missions that were established required cattle for support, but the sparse growth of grass over much of Arizona severely limited livestock operations. Around 1750 efforts at organizing missions in the area south of Tucson met with some success. A presidio was established at Tubac in 1752 to protect the missions and a small group of settlers who had moved in from New Spain bringing with them, of course, their cattle. This led to some increase in the activity of the missions and stock raising. Cattle did reasonably well in the area, and in 1775 Tubac was able to supply livestock to found new missions in California. This was the area's major contribution in establishing cattle in the outer reaches of New Spain.

Throughout the period of Spanish dominion, Arizona received little attention and was never more than a military outpost with a few scattered missions and some settlers. At the opening of the nineteenth century, there were only a few hundred Spaniards in the area including soldiers, missionaries, and settlers. Cattle ranching was the main pursuit of the inhabitants, but the range was not fully utilized until many years later.

California   Colonizing efforts north of the Rio Grande had been a disappointment to the Spaniards. California was far from the seat of authority in Mexico City, and it was difficult to obtain authorization and funds for expeditions to an area in which there was little interest. It was again the forays of other Europeans into lands Spain claimed for her own that sparked her move into California.

The middle of the eighteenth century saw the Russians coming down the coast from Alaska, waiting for an opportune time to come ashore. There were also indications of British interest in lands south of Canada. The viceroyalty in Mexico City at last realized it was time for action in California, and the decision was made to start colonization by moving up from Baja California. Captain Gaspar de Portolá, governor of Baja California, was authorized to recruit settlers and gather livestock to found a mission on the Alta California coast. He directed Fernando Rivera y Moncada, a commander of the presidio at Loreto, to head this expedition to the north. Such livestock as could be obtained was picked up from missions along the route. Portolá later organized a second party and joined that of Rivera before the site of San Diego was reached.

The peninsula was well separated from the prosperous areas of New Spain, and cattle had not increased satisfactorily on the desert lands. Since the missions were reluctant to give up their stock, only 200 head of cattle were collected. These were trailed to the site of San Diego, where the expedition met the survivors of 126 settlers who, coming by ship, had preceded the land party. Sickness had taken a severe toll of both passengers and crew. The mission of San Diego was founded in July, 1769, after the two expeditions joined. The first Spanish cattle had now reached California.

Because the Indian population proved unfriendly, there was difficulty in holding the settlement together. Continued forays by Indians drove the herdsmen back to the mission for protection, and many cattle were stolen. At one point the harassment became so serious that the men tried to force an abandonment. Eventually some success was achieved in converting the Indians to the Catholic faith, and they were enlisted in caring for the cattle. A presidio was founded at Monterey in 1770. During the next few years other missions were founded and stocked with cattle, one as far north as San Francisco.

The second significant movement of livestock to California consisted of 355 head of cattle and 695 horses and mules, which were included in the expedition that Juan Bautista de Anza led to found the presidio and mission at San Francisco. De Anza, captain at Tubac, and later governor of New Mexico, started from Sonora, enlarged his party and picked up most of the cattle in Tubac. When Monterey was reached, de Anza turned back, but the expedition continued on and arrived at the site of San Francisco in September, 1776. This was by far the most important introduction of Spanish cattle to California as the herd arrived nearly intact.

During the next fifty years the missions greatly expanded and were the dominant influence in the Californian way of life. There were few private ranching operations at this time, as the Church was in full control of the economy and brooked no competition. There was little change initially in this pattern when in 1821 New Spain gained her independence from the mother country and became "Mexico." Under Mexican authority, the dominance of the Church in California did eventually come to an end. Talk of secularization of the missions had begun even before independence but no action was taken until 1834. In that year a law limiting the power of the missions was passed. A year later the fathers lost control over their Indian charges, and the prosperous days of the missions ended.

Large ranches took over the establishments and the livestock of the friars. Stock raising continued to provide the main support of the community, but was now the child of private enterprise and not the Church. Political maneuvering in gaining control of the herds of the missions played an unsavory part in establishing the new order. By the time the area that included California had been ceded to the United States, the influence of the Church in the economic life of the community had practically ceased.

The cattle population of California until the middle of the nineteenth century consisted of the pure descendants of the Spanish cattle which had been brought up from New Spain during the previous century. They were known as "Spanish" cattle.

Hawaii   The northern trail of cattle out of New Spain did not come to an end in California. The terminals were out in the Pacific and up in Oregon Territory which included the present states of Washington and Oregon.

George Vancouver, the British navigator, took five cows and a bull he had purchased in San Francisco to the island of Hawaii in 1793. These he presented to the king, evidently to curry a bit of royal favor; Vancouver at the time was endeavoring to have the islands annexed by Britain. A few more head were sent over in subsequent shipments—all to join the king's herd. These Criollos had survived a century in arid Baja California and Arizona and had then enjoyed two decades on good California grass.

On reaching Hawaii, the king's charges were allowed to

run free under a royal taboo against any molestation for 25 years. During this period they multiplied to the point where they became a major nuisance to the farming community on the island. The taboo was finally relaxed to permit a limited harvest of animals for hides and beef. The herd continued to increase, however, and spread out to the uncultivated land in the mountains where they remained wild. Locally, these descendants of the Criollo were called "Hawaiian Wild Cattle."

Oregon Territory was the northernmost point on the continent reached by the Spanish cattle. It was also the one place where they were bred out soon after arrival. Settlement of the territory began in 1834, and during the first years the area was woefully short of breeding stock. It was a long and difficult trail from the Missouri River origin of the wagon trains to the Willamette Valley of Oregon Territory. Marauding Indians with a taste for beef were often encountered. The first wagon trains were not organized to trail cattle, and the only animals they arrived with were their ox teams and an occasional milk cow. Breeding stock, of necessity, were left back east.

Oregon
Territory

A few of the early settlers became aware of the herds to the south, and in 1836 a drive was organized to trail cattle up from California. A total of 729 head were purchased around San Diego and headed for the Willamette Valley. There were continual Indian attacks as the herd moved north, and a hundred head were lost. Those that arrived were a most welcome addition to the small numbers in Oregon. This paved the way for other drives that soon followed.

In 1841 there was again a major drive from California totaling 600 head, and in 1843 another of 1,250 head. These were the last of the large drives. Many wagon trains were now coming in from the east, and the trail bosses had learned the technique of getting cattle through Indian country. Wagons were drawn into a circle at night, and the cattle were driven inside where they were protected. As it was now less hazardous to bring cattle out from the east, the drives from California ended. Shorthorn cattle were becoming available and were much preferred to the Criollo.

For more than a decade the Spanish cattle from California played a vital part in founding the breeding herds in Oregon. They supplied a cow base which permitted a much more rapid increase in the cattle population than would have been possible relying only on cows from the east. The Spanish cow later furnished the same service on the ranges from Texas north to Colorado, Wyoming, Montana, and even into Canada. All trace of the Spanish cattle in Oregon was bred out during the latter half of the nineteenth century.

Texas
As in California, it was the threatened encroachment of outsiders on their domain that led the Spaniards into East Texas. The first scare of French activity in Texas was met by sending out two expeditions that reached as far as the Neches River. Two small missions were established there in 1690. These certainly had cattle which were either abandoned or left to the Indian converts when, three years later, the withdrawal back to New Spain was ordered. Danger of the French at this time was considered to be a false alarm. Spain had gained nothing in the way of a return from her enterprises north of the Rio Grande and, to cut expense, abandoned East Texas. Whether the cattle left in the Neches area survived does not appear to have been recorded. They may have been cared for by the Indians for a while and could well have been the start of a wild cattle population.

In the early 1700's the attention of the viceroyalty in Mexico City was again directed toward the nondefined border area between French-occupied Louisiana and East Texas. French explorers had been wandering into the area and establishing trading posts, stocked with goods from New Orleans. The Spaniards, deciding that their presence needed to be displayed, now made a more determined effort at settlement. The established pattern of sending in military forces to build presidios, accompanied by friars to found missions, was followed. Settlers then followed in their wake. Cattle, as usual, were taken along, and when the Indians could be subdued and entered the missions, they were trained in the care of livestock.

Six small missions were founded north of Beaumont in 1716. This was followed by the establishment of the presidio and mission in San Antonio in 1718. The founding expedition had "7 droves of mules, 548 horses and an undesignated number of cattle and goats." (Bannon, *Spanish Borderlands Frontier*) The number of horses is given exactly, but the cattle are only mentioned as being present. All of these founding expeditions had worked their way north from Coahuila Province and crossed the Rio Grande at what is now Eagle Pass.

Over the next few years there was a modest expansion of mission activity concentrated mainly around San Antonio. Goliad was founded in 1749. All the missions had cattle, horses, and usually sheep and goats. Their herds increased rapidly, often to the extent that control was lost; cattle frequently strayed and became wild.

Settlements followed the missions; some large haciendas were established, but there were no important colonizing movements. The Spanish advance had been directed solely toward establishing outposts to prevent French encroachment on Texas soil. When Spain obtained title to Louisiana in 1762, the need for such measures vanished. Texas lands were at the far reaches of the Spanish dominion of New Spain and attracted little attention during the last years of the eighteenth century. The major settlements of San Antonio, Goliad, and across the river at El Paso del Norte continued their existence but with little attempt at expansion. The area further south between the Nueces River and the Rio Grande, however, had received sufficient protection from the Indians so that the haciendas were able to hold their own without much government help. Here was the cradle of the Texas cattle industry.

The cattle population increased in an unaccountable manner. Before the opening of the nineteenth century, some missions counted their herds in the thousands. The large mission near Goliad claimed 40,000 and several others over 10,000.

Alongside the missions, ranchers had acquired large cattle holdings under the protection provided against the Indians. Those were the days when the famed w'ld cattle of Texas began to multiply and spread through the wide range north of the Rio Grande.

The Spanish cattle that grazed the plains of southwest Texas at the end of the eighteenth century—the open-range herds of the missions and early cattlemen as well as the wild cattle back in the mesquite—had remained uncontaminated by any other bovine influence. They were called "Spanish" cattle, and a few decades later became known as "Mexican" cattle. To the northern cattlemen they became "Texas" cattle or "Longhorns." These formed the base stock of the nostalgic Longhorn—a Criollo carrying a minor influence of the Native American cattle—which made its appearance in Texas during the first half of the nineteenth century.

Part III. END OF THE TRAIL

The Spanish cow reached her zenith in the Western Hemisphere in the first decades of the nineteenth century. Eastern United States, Canada, and Brazil were the only areas in the hemisphere where other kinds of cattle existed. The progeny of the few hundred head of Spanish cattle which had been brought to the island of Hispaniola four centuries earlier had spread south to the Argentine, north to Oregon Territory, and across both continents. The line of descent of this vast population is well founded. Even today the similarity of the few remaining descendants of the old Spanish cattle, widely scattered throughout the hemisphere is inescapable. The year of 1512 was set as the time when the foundation herd on Hispaniola that populated Spanish America with cattle was closed—variation of a few years from this date is of no moment.

During early colonial days there was the occasional shipment of a parcel of cattle from Spain or the Canary Islands to the Indies and the mainland. The "selected cows" that went to Puerto Rico in 1541 and the 24 head of fighting cattle that reached Mexico in 1552 have been mentioned. An exhaustive search of the record would undoubtedly reveal others. Later there were also odd movements of cattle to the New World that did not originate on the Iberian Peninsula. That some cattle from France accompanied Iberville to Dauphin Island in 1699 has been noted, also the movement of French Canadian cattle to Louisiana in the last half of the eighteenth century. There may well have been other minor bovine infusions to Spanish America before the invasion of the northern European cattle and the Zebu began in the last half of the nineteenth century. All such increments of cattle, however, were small in number as compared with the populations to which they were introduced and eventually absorbed. The genetic composition of the Spanish cattle which covered Spain's empire in America in 1800 could not have been significantly altered by these additions.

The original Spanish cow had now become the Criollo in all Latin America—transformed into several different cattle types. The small, half-wild beast of the Venezuelan llanos, the very fair milk cow of the Cauca Valley farmer in Colombia, the rougher bull on the high plateau of Mexico—all were line descendants of the original Spanish cattle; all were Criollos. Each of these highly differentiated cattle types evolved from the one small gene pool.

Differences in size, conformation, productivity, and disposition had been imposed by a wide range in environments and varied levels of management. In Europe such contrasting cattle types would have been recognized by breed names. In Latin America only occasionally was the Criollo recognized by a distinguishing name such as the Blanco Orejinegro in Colombia or the Llanero of Venezuela. After foreign cattle were introduced, those carrying a hump were known as Zebus; the northern European types were called by their breed names.

In the United States, the descendants of the Spanish cattle acquired various designations. In Florida they were just "scrub" cattle. Here, Scrub will be capitalized when referring to the pure descendants. In Louisiana, Texas, and California, where the stockmen already had some familiarity with other kinds of cattle, those from south of the Rio Grande continued to be known as "Spanish" or "Mexican" cattle.

There are areas where the pure Criollo has completely

disappeared, others where small groups can still be seen showing typical Criollo characteristics. In a few countries successful
efforts are being made to perpetuate the Criollo.

At the opening of the nineteenth century there were three
basic cattle populations in the Western Hemisphere, all separated from each other as indicated in Figure 2.

1. The *Criollo*, descendants of the original Spanish cattle that
   were introduced to the Indies.

2. The *Native American cattle*, confined to the settled areas from
   the eastern seaboard of the United States westward to the
   Missouri River. Their genetic counterparts were in eastern
   Canada. Obviously, there were no "native" American cattle.
   The term was used by mid-nineteenth century livestock writers
   in a generic sense in referring to the mixed progeny of the
   different types of European cattle which had been brought
   to the British colonies in early colonial days. Cattle of recognized breeds did not play an important part in the United
   States livestock industry until after the Civil War.

3. The descendants of the *Portuguese cattle* that had been introduced to Brazil in the first half of the sixteenth century. These
   Brazilian cattle were well separated from the Spanish cattle
   by geographic barriers and large unpopulated areas. In only
   the few instances which have been mentioned had even minimal mixing of Spanish and Portuguese cattle occurred.

The breeding out of the pure Criollo began around 1800
in the United States when the early settlers from the southern
states started moving into Louisiana, and a little later into
Texas before that Republic joined the Union. Even the fore-
runners of those who pioneered into these areas often brought
along their milk cow and traveled by ox team.

The Spanish cattle which reached Oregon in the 1830's
were bred out, as we have seen, soon after arrival. In the rest
of the hemisphere, the Criollo remained unmixed with other
cattle until the second half of the nineteenth century. The cattle
types which were then introduced into the different Criollo
populations were:

LEGEND

Spanish Cattle

Native American Cattle
(Northern European Pre-breed Cattle)

BRAZIL
(Portuguese cattle)

N

Fig. 2. Distribution of the Founding Cattle Types in the Western Hemisphere, 1800

1. The breeds which had recently been developed in Britain and northern Europe. During the preceding 60 years, local cattle types in Europe had reached the status of "breeds" as sponsored by breed societies and recorded in herdbooks. The British beef breeds, *Shorthorn, Hereford,* and later *Angus,* were used extensively in breeding out the Criollo in the temperate zones of the hemisphere. The dairy breeds, *Jersey, Ayrshire,* and the *Holland Black and White* (this one had no herdbook at the time), played a later and smaller part.

2. The *Zebu* breeds of India. These cattle were first introduced into Jamaica, then into Brazil, and soon reached nearly all tropical and subtropical areas of the hemisphere. New Zebu breeds were later developed from the Indian breeds, particularly the American Brahman in the United States and the Indo-Brazil in Brazil. After World War II, the American Brahman practically replaced the Indian Zebu in breeding out the Criollo throughout Latin America, except in Brazil.

The Criollo was first bred out in the United States and the Argentine; Colombia and Cuba today have the largest numbers left; in several countries a few representatives remain in unorganized herds. The decline of the Criollo will now be followed through the years down to the position it occupies today.

South America

Uncounted millions of descendants of the old Spanish cattle had spread over South America by the middle of the nineteenth century. In value they far exceeded all the treasure Spain had obtained from the mines of the Andes. The annual offtake at that time produced more income than any other industry on the continent. But over the next hundred years the great Criollo herds of the llanos and the pampas, as well as the milk cow and ox of the small farmer, were to be almost entirely replaced by other bovine types.

Argentina Spanish cattle which had been brought to the northern Argentine lowlands in the 1570's had spread to the south by the end of the century and had increased beyond human need. The beef trade to the mines of the Andes had made little impression on their numbers. When silver lodes began to play out toward the end of the seventeenth century, trade to the mining districts declined rapidly. Buenos Aires replaced Lima as the supply point for northern Argentina. The pack trains which had moved across the mountains with supplies for the settlements and then returned with grain and hides were supplanted by long strings of oxen hauling wagons from the La Plata to the interior. On the open range both the wild cattle and the holdings on the large land grants continued to multiply for another two hundred years.

During the last quarter of the eighteenth century Buenos Aires became the center of Spanish commerce below the equator. Cattle began to play a more important part in the economy. In 1778 the custom records showed that 150,000 hides were exported; five years later this had increased to 1,400,000. Dried beef next became an article for export as well as for home consumption.

The systematic processing of cattle for beef began in the early 1800's. The saladero, a rudimentary packing plant that had been developed in Uruguay, was introduced around Buenos Aires in 1810. The operation began with a supply of cattle held in a corral adjacent to the other facilities. After killing, the hide was removed and salted for shipment. The flesh was then cut from the carcass in strips, soaked in brine, and salted as it was packed in barrels for shipping. The fat was rendered from inedible portions and marketed as tallow.

The saladero was the beginning of an organized cattle industry in Argentina. It served as the means to utilize the entire carcass until long after the first British packing plant was built in 1882. In remote areas many cattle were still killed for their hides because of lack of transportation, but by 1860 forty per cent of all animals slaughtered were processed in the saladero.

Shipments of live cattle to Britain began in the 1870's, but met with poor acceptance. The lean, tough meat of the aged Criollo animal held no appeal to the British palate. The inferior quality of fresh beef, along with the saturation of the market for salted beef, led to the introduction of the purebred animal and the breeding out of the Criollo.

A few British purebred bulls had been brought in and used on Criollo herds in the 1840's. The fatter carcass and softer meat was discounted by the saladero operator who still preferred the lean, dry flesh of the Criollo. It was the development of the market for fresh beef, either live or refrigerated, that brought the British breeds to the Argentine to stay.

British companies, with large ranching operations in the pampas, began to use Shorthorn bulls extensively in their Criollo herds in the 1870's, and were soon followed in this practice by the Argentine stockman. Next came the Hereford and then Angus sires. In a few generations the pampas were dotted with crossbreds of the British cattle. Purebred herds were established and availability of bulls of the British breeds increased. The crossbred British-Criollo females were bred back to British bulls. This upgrading continued and became the general practice. After a few generations, all trace of the Spanish strain disappeared. By the middle of the twentieth century, Shorthorn, Hereford, and Angus herds dominated the pampas.

In the 1880's the refrigerated ship came into use, increasing the availability of Argentine beef to the British consumer. Large abattoirs were built around Buenos Aires to meet the British standard for fresh beef. After the turn of the twentieth century, packers from the United States moved in, built their own plants, and further expanded the market, particularly for canned meat. The Argentine economy had been founded largely on the cattle industry, the base of which was the Criollo cow.

A few representatives of the British dairy breeds were imported as early as 1866. Eventually large dairies were established around the cities, built up from a Criollo base. This also contributed to the disappearance of the old Spanish cattle.

In semitropical northern Argentina the Zebu bull was im-   *95*

The last of a vanished race: A pen of four-year-old Zebu-Criollo crossbred steers from northern Argentina consigned to the Buenos Aires stockyards, 1964.

ported from Brazil and used in Criollo herds. The Zebu-Criollo cross, however, played a minor part in the cattle industry during the period of transition to the British breeds and eventually disappeared.

Argentina and Uruguay were the last regions of South America to see the Spanish cattle established; they were the first to see them disappear. Criollo herds had founded the

Argentine cattle industry, furnishing the base which allowed conversion to the British breeds in less than a hundred years. By the mid-1960's a steer showing any Criollo influence was almost unknown in the Buenos Aires stockyards.

The cattle industry in Uruguay had its inception in the eighteenth century when the roaming gauchos began to slaughter the wild cattle for their hides. Commercial use of the flesh began in the 1780's when the first saladeros were built along the banks of the Rio de la Plata. For nearly three decades Uruguay had the advantage of this factory-type operation for utilization of the beef carcass before it was introduced into Argentina. The lean flesh of the Criollo cattle was the ideal raw material for the saladero. There was no incentive to bring in the British breeds until the packing plant reached Uruguay. Liebigs were manufacturing their famous Beef Extract from Uruguayan cattle in the 1860's, but here again, the Criollo carcass was satisfactory raw material. **Uruguay**

What had appeared to be an unfailing supply of wild cattle soon began to disappear. Breeding and control of cattle to ensure a supply of raw material to the saladeros became necessary. Estancias, founded on large land grants, grew up on the east side of the Rio de la Plata. The establishment of ranching operations followed much the same pattern as it did in the United States at about this time. Settlers moved out onto the virgin grasslands with oxen-drawn covered wagons not too different from those the Alabama farmers drove into Texas in the 1830's. Ranching then began to take on a modern aspect, pastures were fenced, and wild cattle were to be found only in remote, unpopulated areas. The entire cattle population remained pure Criollo until well into the twentieth century.

When the British purebred reached Uruguay, the upgrading process followed the same pattern as in Argentina with the exception that the Hereford remained the favored breed. Large estancias established their own purebred herds in order to obtain a ready supply of bulls. The breeding out process, how-

Mixed criollo cattle from a small herd in central Uruguay on the trail to the railhead for shipment to the market in Montevideo, 1964.

A group of cows crossbred from a Criollo base. The herd in the Chaco had been gathered for heat detection and was one of the first in the country to be bred artificially. West of Concepción, Paraguay, 1964.

ever, was slow. In the mid-1960's, crossbred criollo could still be seen in herds of less progressive owners, but the pure Criollo had disappeared.

Paraguay

The breeding out of the Spanish cattle in Paraguay occurred later and was not as rapid nor as thorough as in Argentina and Uruguay. British cattle companies, inspired by their success in the Argentine, began to establish large ranching operations in the early 1900's. The Criollo cattle, under better management, did exceptionally well initially. Herds of 10,000 head and upward became common. Three large packing plants were built to process beef products for export.

Hereford bulls were then brought in to use on the Criollo cows to obtain a better carcass. The Criollo had become well adapted to the heat, the tick, and the coarse grasses of the Chaco, the wet savannah west of the Paraguay River; but the pioneering was too rough for the British purebred. Heat stress, even in the first cross on the Spanish cattle, was debilitating and increased in the second and third crosses. The screw worm took a dreadful toll. Disease and calving losses were such a serious handicap to ranchers attempting to establish Hereford herds that the better operators compromised with a mixed criollo carrying less Hereford influence.

Later in the twentieth century, Zebu-type bulls were introduced. These were used on both the pure Criollo and the Hereford-Criollo crosses. Resistance to insects and heat stress was substantially increased, and better weights were obtained on the steers that went to slaughter. The hybrid vigor of the crossbred also contributed to the improvement. American Brahman and Santa Gertrudis bulls were brought in after World War II and were extensively used. Eventually some herds of what would be called grade Hereford, but carrying a degree of Zebu and Criollo influence, were established.

The pure Criollo was gone in Paraguay by the close of the 1960's. By that time even the Zebu-Hereford-Criollo crosses showed little influence of the old Spanish cattle.

The Zebu was eventually brought to the Guianas for upgrading the Criollo. (Guyana, having been stocked mostly with Portuguese cattle, is outside the Spanish cattle picture, but crossing with the Zebu also occurred there.) In Surinam, little attention was given to stock raising on the Dutch plantations. With water transportation for cane, there was little need for the draft ox. Zebu breeding stock was not brought in until 1948. Cross-breeding then progressed rapidly, and in a few generations the only Criollo left were the small farmer's milk cow and his span of oxen.

Guianas

The French appear to have been interested in the Guianas mainly to retain a foothold on the continent. In French Guiana sugar production was finally developed, and sizable plantations flourished during the first half of the twentieth century. Criollo cattle were used for draft until the sugar economy deteriorated after World War II. There was crossing with the Zebu but not as extensively as in Surinam. With the abandonment of the cane fields, most of the cattle disappeared. The few thousand head remaining today generally show Zebu influence although occasionally an almost pure Criollo cow can be seen on a small farm.

Western South America below Colombia never saw the large cattle developments that took place to the east of the Andes. Neither the arid coastal plains nor the altiplano of the Andes had environments to encourage the cattle industry. The settlers who followed the conquerors into Peru and Ecuador had their milk cow and ox team, and their progeny account for the Criollo that has survived down to recent time. Large ranching operations never developed, although a few moderate-sized holdings were established in the areas where the coastal plain widens. Such areas as the Piura River drainage in northern Peru, the Costa surrounding Guayaquil in Ecuador, and the central valley in Chile all became stocked with the Criollo to the limit of their carrying capacity. The cattle served the needs of the inhabitants in the agricultural areas but beyond

Chile, Ecuador, Peru

In the high altitude of the sierra, the Criollo is a small animal and frequently lacks the horn of typically Spanish cattle.

A Criollo bull of nearly pure Spanish descent in the holding yard of the Quito abattoir. Ecuador, 1966.

A Criollo cow in the Chillo Valley—a chance survivor in a herd where Hereford bulls had been introduced. This black and white color pattern is fairly common in the sierra Criollo. Ecuador, 1966.

American Brahman-Criollo crossbred heifers in a herd that was being upgraded at an estancia on the Duale River in the Costa, Ecuador, 1966.

107

this did not enter into the general economy. Larger towns and cities were as dependent on the importation of beef as the early mining districts of the Andes had been.

When the northern European purebred became available in the 1900's, the dairy farmers of Chile were quick to import the European dairy breeds—Friesian bulls from Holland, the Red and White cattle from Germany, and later the Holstein-Friesian from the United States. The pure Spanish cattle were completely bred out in a few generations. Even the mixed criollos disappeared from Chile.

In Ecuador the Criollo became well acclimated to the high altitude of the sierra. The small farmer has had his ox team and milk cow since colonial times. The few moderate-sized herds that were held on good grass areas were being bred up to Hereford and other northern European breeds in the 1960's. Pure Criollo cattle are still held on small farms and are occasionally seen in herds in the process of being upgraded. But these will certainly be gone in a few generations.

For years the dairymen around Quito have imported bulls of the dairy breeds to upgrade their Criollo herds, the Holstein-Friesian from the United States being preferred. Purebred females were also imported for the establishment of registered herds. No commercial dairy herd now shows any Criollo influence.

On the Costa, the humid coastal plain of Ecuador, the pure descendants of Spanish cattle were raised mainly for draft until well into the twentieth century. There had been some crossing with Zebu bulls from Brazil after the turn of the century, but the herds remained largely Criollo for many years. The 1930's saw the American Brahman bulls being used for crossbreeding. A number of sizable ranching operations were established. Breeding for a beef-type animal became the common practice; herds were upgraded American Brahman cattle or criollo crossbreds. No trace of the Costa Criollo cow is left to enable a comparison with her small sister on the sierra.

Criollo herds had been maintained since colonial times in Peru on the northern coastal plain and the lowlands of the

Piura River drainage. Cattle holdings in the rest of the country were small, usually only a few cows and the ox team of a subsistence farmer in the mountains.

Crossbreeding to Gyr bulls imported from Brazil started in the Piura area. After World War II, the American Brahman began to be used; also a few Brown Swiss herds were established to provide bulls for upgrading the Criollo. As the use of draft animals decreased, emphasis turned to beef production, and within a few generations the Criollo was replaced by grade American Brahman and Brown Swiss cattle. Only an occasional animal showing a marked Criollo influence could be seen in the 1960's.

The dairy herds around Lima are of northern European breeds which have to a large extent been bred up from a Criollo base. There is a sprinkling of imported stock in which the Holstein-Friesian predominates.

Cattle have never been an important element in the Bolivian economy. Development of the country since the days of the conquest has been in the mining regions of the altiplano, the high plateau area in the Andes. Early commerce followed the old Inca military road north to the port of San Miguel, and south into Argentina. Rail and air transportation have replaced the ancient highway, but the mines, now producing tin instead of silver, are still the lifeblood of the country.

Bolivia

The few cattle now in the altiplano are the descendants of early Spanish cattle that became adapted through natural selection to the high altitude and the scant vegetation. The mines in colonial days depended for meat on cattle trailed up from the northern pampas of the Argentine; today fresh beef is flown in from the Beni, the Bolivian lowlands of the Amazon River.

The cow and ox span of the altiplano subsistence farmer today carry a varying but usually small degree of influence of the European dairy breeds. This results from past efforts to establish Holstein-Friesian or Brown Swiss herds for milk

109

Criollo ox span of a subsistence farmer on the altiplano south of La Paz. These animals weighed around 700 pounds. The small size is the result of countless generations born and raised on the sparse vegetation at the high elevation. Bolivia, 1969.

production in the vicinity of La Paz. The 12,000-foot altitude was severely debilitating to the European cattle, which factor, along with the effect of land reform efforts, resulted in failure to establish a milk shed. An occasional individual can still be seen, however, that is essentially pure Criollo. These small, hardy beasts show the remarkable ability of the old Spanish stock to adjust to any environment in which they are placed.

110

Four centuries of natural selection have produced a type of cattle adapted to altitudes over 12,000 feet. Individual cattle holdings are limited to a few head, raised for use as draft animals and milk for household use. Only when no longer useful, or because of the owner's urgent need for cash, does one go to the abattoir.

The Criollo cow of the altiplano weighs perhaps 650 pounds, and the bulls 750 pounds. The color displays various shades of brown, often with some white on the flanks and underside. The hair is thick and much longer than at low elevations. The horns, widespread and large for the body size, show the Spanish influence.

The few Criollos remaining on the altiplano do not appear to be in any imminent danger of extinction. They are essential to the farmer's livelihood and will continue in his care until his economic status changes.

The main cattle country in Bolivia is the lowlands of the upper Amazon basin, a region which until modern times lacked communication with the prosperous mining settlements high in the Andes. After railroads were built, cattle for slaughter came in from Peru to the centers of population in the mountains. It was not until the advent of the airplane that the cattle-raising potential of the oriente became a factor in the Bolivian economy.

The Spanish cattle which were brought from Asunción in 1561 to the lowlands east of the Andes continued there with no mixing with other cattle until well into the twentieth century. There may have been some contact with Brazilian cattle from across the border in later years. If so, the influence on the Criollo could not have been significant as the Brazilian herds were descendants of Portuguese cattle which were close relatives of the old Spanish cattle.

The region known as the Beni of northeastern Bolivia is a water-logged forest in the rainy season, interspersed with small savannahs at slightly higher elevations. It is not an ideal cattle country but the old Spanish animals adapted to the environment. Ample grass, although deficient in protein and

These Criollo cattle were brought to the La Paz abattoir holding yard from small altiplano farms:

The bull shows a strong Criollo influence although the head and sheath are not typical. Bolivia, 1969.

112

The thick and long hair on this cow is due to the cold temperatures of the high altitude. Bolivia, 1969.

These are typical Criollo cattle seen in the Beni—the cattle country of Bolivia—but are much larger than their relatives on the altiplano:

Showing the black and white Criollo pattern that is quite common in cattle of the Beni, this bull also has the black ears of the Colombian Blanco Orejinegro. Bolivia, 1969.

The cow displays the typical Jersey-tan color of the Criollo. Bolivia, 1969.

116 *These crossbred cows, the progeny of Nellore bulls on Criollo females, are the result of a controlled breeding program in the Beni. Bolivia, 1969.*

minerals, was always available. A hot, humid climate and biting insects had to be endured and adjusted to. The Criollo cow which emerged from these conditions was well acclimated and productive and did not differ markedly in appearance from her sister that evolved in more hospitable environments.

Rather primitive ranching operations became established in the Beni, but cattle raising attained little economic importance because of lack of a market. In the rainy season, the only transportation was by boat along the numerous waterways. The only outside market of the oriente was a clandestine trade with Brazil. There was little outlet even for hides, the traditional source of income in areas where cattle are in surplus and there is no market for beef. Cattle increased in number far beyond any conceivable demand.

In the higher, more open country around Santa Cruz, south of the Beni, stock raising followed much the same pattern. A crop-farming community afforded a somewhat better market, but a surplus of cattle existed here for years.

The heavy demand for beef during World War II eventually reached into the Beni for cattle. The Bolivian national herd went from two million head to a few hundred thousand before the buying spree was over. The heavy movement had been by air and proved the advantages of this mode of transportation to the growers in the Beni. After the war, availability of surplus army planes made it possible to continue the shipments of beef by air and thus reach the market in the mining districts. The era of prosperity that began with the wartime sales now enabled the Bolivian cattleman to purchase bulls for crossbreeding. Zebu bulls, mostly Nellore and Gyr from Brazil, were brought across the border and put on the Criollo herds. Crossbred Zebu-Criollo bulls were also widely used. By 1970 the Criollo had nearly disappeared, replaced by crossbred cattle of varying percentages.

One island of pure Criollo cattle still remains in the Beni. The Espiritu Ranch, running several thousand head of criollo cattle having a heavy Zebu influence, still maintained in the late 1960's a pure Criollo nucleus of 500 head. From this herd

Criollo bulls were produced for a program designed to maintain a fixed Criollo influence in the crossbred herd. This holding of Criollos was among the largest in the hands of one owner in South America at that time.

Government experiment stations in a few Latin American countries are endeavoring to maintain pure foundations of some of the old Criollo breeds that trace back to the elementary selection practices of colonial husbandmen. The Espiritu herd is unique in that it stems from a population which had been only under natural selection until recent years.

The Criollo of the Bolivian lowlands is of medium size, cows averaging 900 pounds and bulls up to 1,200 pounds. The predominant colors are Jersey-tan, or black and white mottled. Horns are quite large and usually on the cow have the typical Spanish shape—wide and upthrust. The horns on the bulls are thicker and rather short. The thin tail with a short brush, high tail stock, narrow face, and dark coloring around the eyes are all true to the Criollo type.

These two survivors, the near-pure Criollo of the altiplano and her larger, better muscled sister in the lowlands, show the wide variation that three centuries of natural selection in the different environments produced from the small genetic base of the original Spanish cattle. The altiplano cow has probably descended from cattle dropped off from a herd headed for Argentina in the late sixteenth century. The Criollo in the Beni descended from the same cattle that were brought over the mountains, down to the northern pampas, and then to Santa Cruz only a few generations later.

In 1810, when Simón Bolívar was taking his place as leader **Venezuela** in the colonial independence movement, the cattle population of Venezuela was around 4,500,000 head. These were all true descendants of Spanish cattle, as there had been no opportunity for mixture with any other type. Bolívar, a native Venezuelan, rose in stature above all the Latin American liberators. He nearly established a United States of South America, riding to

Typical Llanero cows on the Venezuelan llanos at the Hacienda Mara de Barbara, Venezuela, 1966.

glory on campaigns financed by the Criollo herds of the llanos. The hides and carcasses of those beasts bought his armament and fed his armies until, by 1823, their numbers had been reduced to 256,000 head.

The Criollo cow, however, was by this time well acclimated. Those that escaped the roundups were the hardiest of the lot, and in only ten years the cattle population grew back to 2,400,000 head. (White, *Cattle Raising*) While cattle numbers suffered many other wide fluctuations during the next eighty years, no significant introductions of other types occurred.

Guzmán Blanco, one of the long line of Venezuelan dictators, was a cattle-minded ruler who tried to bring better management to the llanos. He is reported to have introduced better types of cattle to the region in the 1870's. Quite probably

120

the improved stock were Criollos from the better growing areas around Lake Maracaibo or Valencia. It is doubtful if any of them made much imprint on the Criollo of the llanos.

Throughout the remainder of the nineteenth century, periods of revolutionary activity followed at frequent intervals. The Criollo cattle on the llanos were a self-perpetuating source of income for the warring factions. Stockraising was largely a matter of accepting the natural increase as ranching operations lacked organization. Cattle were always something that an armed force could draw on to supply food or cash. The right of ownership which resided in a city-dwelling land grantee could be ignored with immunity. During periods of relative quiet, the llanos herds increased as the customary offtake was less than their growth. When war broke out, they   *121*

122 *Upgraded American Brahman-Criollo cattle collected for spraying on a small hacienda near Libertad, Venezuela, 1966.*

would again be decimated. These fluctuations in cattle numbers carried on into the opening years of the twentieth century.

Environmental influences had produced two distinct types of cattle: the small, wild and wiry beast of the llanos and the larger, better fleshed animal that was acclimated to the more hospitable environment of Lake Maricaibo and the coastal highlands. Here there was some selection for desired characteristics in the farmer's animal as will be noted later. The llanos herds, ranging uncontrolled, existed in a nearly wild state and were subject only to natural selection until well into the twentieth century.

Another dictator, Juan Vicente Gómez, who came to power early in the twentieth century, was for many years a dominant factor in the cattle industry. He acquired vast holdings, forcing unfortunate ranchers out of business by imposing confiscatory taxes. The national herd doubled during his regime, much of the increase accruing to himself. Some interest in the British breeds developed among the prominent cattlemen in the last years of Gómez' power but only a few bulls were brought in. There were no large importations of foreign cattle until after World War II. Zebu bulls were then brought in for upgrading the pure Criollo in the Valencia area and around Lake Maracaibo.

The Zebu was followed by purebred Brown Swiss and some Holstein-Friesian breeding stock in the herds of the more progressive dairymen. Thus the breeding out of the Spanish cow began, but progress was slow for the next decade. The large beef herds, which came to be known as Llaneros—cattle of the llanos—ranged in the inaccessible savannahs of the Orinoco River. City-dwelling owners paid little attention to their cattle. Roundups for branding and cutting out animals for slaughter provided the only contact with man. The vast ranges of poor-quality grasses are largely under water during the May–November rainy season and are then dry for most of the remainder of the year.

Nearly three centuries of natural selection on this type of pasture, plagued with biting insects, an unbearably hot climate, and a general lack of breeding control, had resulted in a small, long-legged animal with large, wide upspread horns, and a poor beef conformation, but exceedingly hardy—the Llanero. A mature Llanero cow at seven or eight years of age weighed 650 to 700 pounds, and bulls weighed only a little more. Color was predominantly a solid Jersey-tan although other patterns were seen, particularly a nearly solid black, and mottled black and white. No explanation is apparent of where the Criollo, reported as solid red or brown in the early 1800's, acquired the black, and black-and-white patterns.

In 1950 the cattle in the llanos, two-thirds of the national herd, were the Llaneros—pure Criollos or criollos showing a strong Spanish influence. Importations of Zebu breeding stock then increased rapidly, coming first from Brazil. Eventually the American Brahman became the popular breed. Purebred herds were established to provide bulls to meet the demand for upgrading. Around the cities there was increased use of the northern dairy breeds, mostly Brown Swiss and Holstein-Friesian. One controlled crossbreeding development began at this time in which the Criollo was used as a means of adapting a northern European dairy breed to the Colombian environment. This will be examined later.

By the mid-1960's the pure Llanero was rapidly disappearing. Only an occasional cow reminiscent of the old Criollo could still be seen in the remote areas on the land of a less progressive rancher. Today most of the herds in the more accessible areas of the llanos are upgraded American Brahman. There is some representation of Santa Gertrudis, but this breed never became very popular with the Venezuelan cattleman. In the far reaches of the llanos the cattle are criollo, American Brahman-Criollo crossbreds, still in the process of being upgraded.

The farmers in the more settled area northwest of Lake Maracaibo in Zulia State followed management practices unknown in the llanos. Here some degree of selective breeding was employed before the day of written records and continued down to modern times. Selection for a draft-type animal may

The Limonero, a dairy-type Criollo from the Rio Limón area, was developed for milk production at the Centro de Investigaciónes Agronomicas at Maracay:

This four-year-old bull was awaiting the results of his daughters' production tests before being certified for regular service. Maracay, Venezuela, 1966.

One of the best producers, an eight-year-old cow with a record of 6,600 pounds of milk in one lactation. The photograph was taken during her rest period. Maracay, Venezuela, 1966.

have first been practiced by the farmers, later modified to obtain higher milk production. The Rio Limón area became the principal milkshed of Venezuela. Long before the animal scientist of the experiment station began his daily recordings of individual milk weights, the Zulia dairyman noted the cow that had left the most milk in the pail and saved her daughter for a replacement. Over the course of many bovine generations the cattle population of the Rio Limón area became the "Limonero." Among these pure descendants of the old Spanish cattle, there are herds which average 4,000 pounds of milk per lactation—a phenomenal performance for a native tropical cow.

By 1950, however, the crossing complex had spread to Zulia and other more recent dairy areas, such as the Carora area in Lara State. Brown Swiss and American Brahman bulls were being used in the Limonero herds, particularly those of the small farmers, to get a heavier animal for slaughter. At this point, the Zulia dairymen and the Ministry of Agriculture joined in planning for the preservation of the Limonero breed. The advice of Dr. Jorge de Alba, whose development of the Milking Criollo at Turrialba in Costa Rica will be discussed later, was sought in this endeavor. The broad objective was to develop a higher-producing tropical dairy cow and to preserve the pure Criollo Limonero.

The program was initiated with the selection of 153 cows and 3 bulls from the Rio Limón area, two cows from Carora, and six cows, five heifers, and a bull from Nicaragua. The cattle from the Rio Limón area were selected by Dr. Vladimir Bodisco of the Centro de Investigaciónes Agronomicas at Maracay where the cattle were brought. The females and the bull from Nicaragua were from the Rivas area where Dr. de Alba had obtained many of the animals with which he founded the herd of Milking Criollo in Costa Rica. These cattle were as true descendants of the original Spanish cattle as those from Zulia.

The selected Criollo herd at the experiment station at Maracay was culled to produce an elite herd of "Improved" Criollo, or Limonero, in accordance with modern selection practices. Only the highest-producing cows were kept, and the bulls retained had the proven genetic ability to improve the milk production of their daughters over that of the daughters' dams. In 1967 the best cows in the herd were transferred to the El Laral Experiment Station north of Maracaibo. The number was increased to 300, and the program for higher milk production continued on improved pastures with only a minimum grain supplement fed while milking. The lower-yielding cows were culled and sent to an experiment station in the llanos for crossbreeding experiments with the Zebu.

The Limonero is practically identical in appearance to the Milking Criollo of Costa Rica and closely resembles the Criollo breeds in Colombia. This is particularly true of the Vallecaucano and the Costeño con Cuernos. The hair is short and thin. The face is long with the typical Criollo wrinkles around the eyes. The tail stock is high and prominent; the tail is thin with a short brush. Horns on the cow are wide and upthrust, frequently with a lyre-shaped tip; those on the bull are thicker and not as long. Mature cows weigh 900 to 1,000 pounds in good milking condition, bulls up to 1,600 pounds. The hair color is Jersey-tan, usually with a slight reddish tinge. An occasional animal is seen in a herd of pure Limonero that is solid black with possibly a little white showing around the udder. These fail to meet the standards which have been set for the breed, calling for a "red, bay, yellowish, or brown" color. The appearance of black animals is significant, however, as a black-colored individual is noted occasionally in pure Criollo herds which have not been under any breeding control by man, such as in the llanos or the Central American mountains.

At El Laral, in addition to continuing the work to improve the station's herd of Limoneros, a cooperative program with the local dairymen was inaugurated. Thirteen farms with a total of 1,000 Limonero cows were involved. Artificial insemination initially was employed only in the station herd, but plans were to extend its use to those of the cooperating farmers as the advantages of the procedure could be demonstrated to them.

Individual cows in the best herds have reached a produc-

This bull and cow belonged to the Finca El Playón which had one of the highest-producing Limonero herds (125 head) in the Rio Limón area:

Limonero bull, estimated weight 1,500 pounds (the left horn had been broken in fighting). Carrasquero, Venezuela, 1975.

Limonero cow, estimated weight 975 pounds, is Jersey-tan in color and displays all the typical Criollo characteristics. Carrasquero, Venezuela, 1975.

tion level of 7,000 to 7,500 pounds of milk per lactation. This is for cows in commercial herds, under good management but with no special attention, and shows the genetic potential of the pure Criollo when subjected to modern breed improvement practices for several generations. The purebred cow of the northern European dairy breeds, with uncounted generations of selection for milk production behind her, when transplanted to the tropics and maintained under near-clinical conditions can barely exceed this production of top Limonero cows.

In the decade after the introduction of the European dairy breeds to the milkshed areas of Venezuela, the usual difficulties were encountered that follow a northern breed into the tropics. There was a deterioration in general condition and a material drop in production and fertility. To maintain a dairy herd of the European breeds in condition that would insure good milk production along with satisfactory reproductive ability required special care.

A group of dairymen around the town of Carora became disappointed with the performance of pure Brown Swiss, and set out to establish a tropical milk cow in a new breed. Their objective was to obtain the high productivity of the Brown Swiss and also retain the resistance to tropical health hazards of the Criollo. In 1965, after several generations of this mixed breeding, a 3/8 Criollo–5/8 Brown Swiss cross was selected as the proportions of the two breeds to be stabilized. This hybrid was known as the Tipo-Carora (Carora type).

The color of the new breed was similar to that of the Brown Swiss only lighter in shade. Fertility and health were markedly improved over that of the pure Brown Swiss. Milk production approximated that of Brown Swiss cows which had been maintained under carefully controlled conditions. The Tipo-Carora cow weighed 1,000 pounds, bulls 1,500 pounds.

The success with this crossbred led to increasing the Brown Swiss influence. By 1975 the Tipo-Carora, as represented by the herds around Carora, was considered to be 1/16 Criollo–15/16 Brown Swiss. Evidence of the Criollo influence had practically disappeared; the Tipo-Carora cow had become an upgraded Brown Swiss. The Criollo cow had been the means by which the northern European breed had been acclimated to the tropical environment.

Few attempts have been made to incorporate the Criollo genes in a new breed, and such efforts as have been made have usually been abandoned. The Lucerna—the Criollo-Holstein-Friesian-Shorthorn cross—developed in Colombia is one notable exception. The Romana Red, the Zebu-Criollo cross which was maintained as a breed for a while in the Dominican Republic, is now disappearing.

At the end of the Spanish cattle trail in Venezuela, there is but one survival—the Limonero, the tropical dairy breed of the Rio Limón.

The herds of Colombia continued the line of pure descent from the original Spanish cattle until the last quarter of the nineteenth century. It was then that importation of the European purebreds began. There is a record of seven Hereford bulls being taken to the llanos in 1870. If such an introduction was made, the prospective sires could not long have survived the heat and insects. In 1883 a Holstein-Friesian bull and two cows are said to have been taken to the Antioquia area, and in 1884 two bulls and two pregnant cows of the breed were taken to Medellín. (Patiño, *Plantas y Animales en América*) These introductions may have survived but could not have left much impression on the large Criollo population in the general area at the time. An Angus bull was brought to the Antioquia area in 1886. This individual is involved in one of the many speculations as to the origin of the Romo Sinuano.

At the beginning of the twentieth century, European importations increased. Durham (Shorthorn), Red Poll, a few more Angus bulls, Devon, and Suffolk were brought in, many of which went to the lowlands where their breeding life was probably measured in months. The milk breeds were imported to the dairy areas around Bogotá and to the Cauca Valley. A small representation of Normandy cattle was received from

Colombia

A new breed, the Tipo-Carora, in 1966 was being stabilized at 3/8 Criollo–5/8 Brown Swiss; in recent years the Brown Swiss influence has been increased to 15/16:

This young bull, 3/8 Criollo, 5/8 Brown Swiss, was in service at the Cooperative Artificial Insemination Center, Carora, Venezuela, 1966.

A mature cow in the herd of the Hacienda Los Avangues, Lara Province, Venezuela, 1966.

France. At the higher elevations these northern European breeds did reasonably well and within a few decades upgraded Brown Swiss, Holstein-Friesian, and some Jersey herds furnished much of the milk for the centers of population.

Around 1900 there were several arrivals of Zebu cattle. A few Nellore bulls from Brazil had previously reached the Colombian llanos. There is a report of cattle being received from Hamburg, one bull in 1901 and more in 1907. These animals must have been destined for the circus ring, but were probably picked up by a wandering Colombian cattle enthusiast. Zebu cattle, probably of Brazilian origin, were received in 1908 from Puerto Rico and Texas. There was also a direct shipment from India in 1914. The presence of the cattle tick was noted after Zebu bulls had been placed in breeding herds, and public opinion placed the blame on the humped cattle. Further importations of the Zebu were then prohibited by law. After World War II these restrictions were removed. The American Brahman became popular, and large numbers were shipped in from the United States and placed in the Criollo herds in the llanos. Santa Gertrudis bulls were brought in later but were not widely used.

The upgrading of the Criollo in Colombia has expanded down to the present day, though there were some areas where the practice was slow to take hold. The Cauca Valley and the Sinú were outstanding in this respect. There the Colombian cattleman held on to his Criollos. For many generations he had selected for distinct types, had carefully maintained them through the years, and called them breeds. Eventually, however, the upgrading fever struck the areas where these breeds were held, and they declined rapidly.

In 1975 the Colombian national herd of over 24,000,000 head consisted mostly of purebred and grade Zebu; criollo cattle carrying a strong Zebu influence; and purebred or grade European dairy breeds. There remained, however, a few hundred thousand pure, or nearly pure, representatives of the Criollo breeds, most of which were Blanco Orejinegro. There also existed in isolated areas of the llanos scattered descendants of the old Spanish cattle which had known little domestication. These Criollos, the product of four centuries of natural selection, thrive in the harsh environment of the llanos as no other bovine can.

The isolation provided by the widely varied terrain in Colombia was the dominant factor which permitted the different Criollo breeds to evolve. Low coastal plains, high mountain ranges, large rivers—these in various combinations—placed forbidding limits on transportation until modern road-building equipment became available. In colonial days the settlements were separated by distance and limited transportation. Even as the boundaries of human habitation expanded, natural barriers still held many communities isolated from their neighbors. The people remained segregated and so did their cattle.

Natural selection first developed the type of cattle that was adapted to the environment of the high mountain valleys or in the lowland savannahs. Later, possibly even during colonial days, the husbandmen in some areas made elementary selection for a desired trait: milk productivity, a draft animal, color of hair coat, or such other characteristics as took their fancy.

Thus, the Colombian Criollo breeds were formed. They are generally named for the region of their origin. All, except the Blanco Orejinegro, trace to the small gene pool of Spanish cattle imported to the Indies. There is a blank page in the story of the Blanco Orejinegro which will be discussed later.

In the upper Cauca Valley a Criollo known as the *Vallecaucano* (Horton de Valle, also Arton) became segregated. These are descendants of the Spanish cattle that began arriving in the area in 1537. The early arrivals, and those which followed in the next few decades, were the progenitors of all the cattle in this part of Colombia until modern times. The Vallecaucano have now been nearly bred out by the American Brahman and European dairy breeds. In an effort to preserve the breed, the National University at Palmira has recently acquired a

Villecaucano

137

Specimens of the Vallecaucano breed seen in the National University herd at Palmira. These cattle were selected as typical representatives to perpetuate this strain derived from the old Spanish cattle in the Cauca Valley:

A young bull, the scratches on his side caused by rubbing against brush. Palmira, Colombia, 1975.

This mature cow displays typical Criollo characteristics. Her udder is larger than on Criollos that have not been selected specifically for milk production. Palmira, Colombia, 1975.

small herd of thirty cows and two bulls, as true descendants as could be located.

Although a larger, better-muscled animal, the Vallecaucano is a blood relative of the Casanareno, the small, unpampered cow of the llanos. The dominant color is Jersey-tan although with more of a reddish cast than on the Casanareno. The thin tail with the short brush, high tail stock, wide upspread horns on the cow, short thin hair coat, long muscular face, all typical Criollo traits, are seen in the Vallecaucano. Cows weigh an average of 850 pounds. There were no mature bulls in the university herd at the time it was observed, but the two young bulls would have weighed something over 1,000 pounds. The cows have exceptionally good udders for a Criollo. It is obvious that at some time in the past elementary selection for milk production was practiced. More attention to nutrition then followed. The "Valle de Cauca" in old writings is referred to as a "dairy type."

Natural selection in an environment much more hospitable than the llanos produced the Vallecaucano. At 3,000 to 4,000 feet altitude, the climate is warm temperate, and grasses have fair to good nutritional value. The rainy and dry seasons impose no great problem to cattle husbandry, but the standing plant growth declines in nutritive quality before the rains start.

Chino-Santandereano The local cattlemen's association at Bucaramanga, in the Department of Santander, has segregated a herd of 200 descendants of the old Spanish cattle, known as the Chino-Santandereano. These are the lone survivors of that Criollo breed which farmers in the Lebrija Valley have husbanded for the past four centuries. Nearly all the herds in the region have now been graded up to American Brahman or Brown Swiss, or are in the last stages of such transformation.

In an attempt in 1970 to perpetuate a foundation stock of pure Chino-Santandereano, the association purchased the best examples which could be found. A breeding program was inaugurated to increase numbers in the base herd and to produce pure Chino-Santandereano bulls for the association members. The selection program is based on weaning weights and conformance to type. The final objective is to breed the Chino-Santandereano bulls back on the Zebu-Criollo herds which predominate in the area. This introduces an element of hybrid vigor that results in a heavier and better-fleshed calf.

The Chino-Santandereano was developed at an altitude of 2,500 feet, an environment quite similar to that of the home of the Vallecaucano in the Cauca Valley that lies across two major cordilleras to the west. The color of the two breeds is the same, varying in shade from Jersey-tan to a light red. The Chino-Santandereano is the larger of the two animals, cows averaging 950 pounds and mature bulls around 1,500 pounds. The horns have the same shape but on the Chino-Santandereano are thicker and not as wide nor as long as on the Vallecaucano.

The farmers of the Lebrija Valley are said to have been quite cattle oriented. Certainly a stronger draft animal would have been desirable here where the valleys are narrow and the slopes are steep. Thus a significant degree of selection may have been practiced, both for a large animal and a heavier horn to carry the yoke. There was undoubtedly some selection for milk production.

The Chino-Santandereano and the Vallecaucano display a marked similarity. Four centuries of natural selection, coupled with some breeding control by man, in much the same but widely separated environments, produced these two cattle populations that so closely resemble each other.

San Martín was an early settlement at the western edge of the llanos where the land begins to rise to the eastern cordillera. Here the San Martinero was developed in a wide area around the town, reaching from typical llanos savannahs up into the piedmont area where the grasses are more nutritious. The barrier of the high mountains to the west and the long distances across the llanos to settlements in any other direction accounted for the early isolation of the San Martinero breed.

The first cattle were brought into the area up the Orinoco

San Martinero

A 300-head herd of Chino-Santandereano is maintained at the Fondo Ganadera de Santander in the Lebrija Valley near Bucaramanga. The herd is maintained pure for the production of bulls to be used on herds which have been graded up to a Zebu type:

This five-year-old bull weighed 1,625 pounds. The horns indicate possible selection among his ancestors for a draft type. Lebrija Valley, Colombia, 1975.

The marked similarity of this four-year-old cow to the Vallecaucano illustrates how selection for milk production developed the same dairy-type in widely separated regions. Lebrija Valley, Colombia, 1975.

A small San Martinero herd was moved to the Carimagua Experiment Station on the Meta River in the llanos, east of Villavicencio, Colombia. Their poor condition was a result of a deficiency of nutrients in the llanos grasses. The herd is maintained by the Ministry of Agriculture in an effort to preserve this old breed:

A mature San Martinero bull showing some indication of selection for a draft animal. Carimagua, Colombia, 1975.

This mature cow carries the horn typical to the Andalusian Retinto. Carimagua, Colombia, 1975.

River and to the Meta from the Venezuelan llanos. Natural selection then took over for three centuries, followed in later years by some selection for milking ability as well as for a heavier type for draft. There could have been no mixing with any foreign cattle until the latter part of the nineteenth century.

The San Martinero has a rather rough conformation but is a much better beef type than the Criollo of the llanos. The rather narrow head, thin tail, and thin dewlap are characteristic, as is also the Jersey-tan color shading to reddish tan on many individuals. Lighter color on the legs and underbelly is common. The horns are typically Spanish in shape, slightly thicker than on most Criollos. Mature cows, before the grass dries up, weigh 900 to 1,000 pounds, bulls up to 1,400 pounds.

The neck is heavy with a noticeable thickening over the shoulders of the bull. This suggests selection for a draft type. The moderately muscled hindquarters and rather deep barrel could have come from a few Shorthorn genes, which some writers on Colombian cattle suggest occurred. There appears to be no authentic history of the San Martinero, and its genealogy must be assumed from its appearance. This would indicate that the breed today, while predominantly of Spanish descent, may carry a slight influence of a European beef breed, possibly that of the Shorthorn.

The San Martinero is not as well adapted to the llanos as the Casanareno, the Criollo of that region which will be discussed later. A herd from the piedmont area above San Martín was moved to the Colombian experiment station, Carimagua, in the llanos. Within a year there was a noticeable deterioration in general condition, even on the controlled pastures of the station, which are much better than typical llanos grasses. Except for the Carimagua herd and those at a few outlying fincas, the San Martinero has been bred out by the Zebu, principally the American Brahman. Unless the herd at the station is continued, the breed will disappear.

Costeño con Cuernos   Two Criollo breeds, the Costeño con Cuernos (From the Coast with Horns) and the Romo Sinuano, were developed in the

Sinú, the flat drainage area of the Sinú River in northwestern Colombia. Both could trace to the first cattle landed on South America at the Gulf of Urabá. The Costeño con Cuernos carries horns; the Romo Sinuano is polled; otherwise they are quite similar. There is no authentic history of either breed, and what has been written about them is largely conjecture. Both have been nearly bred out by upgrading to American Brahman.

The Costeño con Cuernos appears to have been selected for the past several decades, first, for draft and milk and, later, for beef traits. Whether the European beef breeds have entered to a minor extent into its composition can only be surmised. Both Shorthorn and Hereford bulls were introduced into Colombia far enough back in time so that their influence could have been a factor in the genetic background of the Costeño. But even today cattle of the temperate zone can be maintained in the Sinú only if given special attention, involving such practices as dipping, immunization from endemic diseases, and special nutrition. Fifty years ago, such procedures were unknown. So it seems doubtful if the British breeds could have been maintained long enough in the region to have had much influence on the Costeño con Cuernos.

The Turipana Experiment Station near Montería has the best representatives of the breed that remain today. These cattle show all the features typical of the pure and improved Criollo. The color is solid and uniform throughout the herd, shading from a near-cherry red to a somewhat darker Jersey-tan than is seen on many Criollos. The thin tail, high tail stock, large, wide upspread horns on both cow and bull, short hair—all are distinctive and uniformly present on all animals. There is no evidence of Zebu breeding. Mature cows weigh 1,000 pounds, bulls 1,400 to 1,500 pounds. Conformation is a good beef type. Writing twenty years ago, de Alba referred to the Costeño con Cuernos as an "excellent dairy type." While the breed appears to have been developed for milk production, today it carries all the attributes of a good milk-beef, dual-purpose animal.

The Costeño con Cuernos herd at the Granja Turipana experiment station in the Sinú was established in an effort to perpetuate this old Colombian breed. First developed as a dairy animal, the Costeño con Cuernos now tends toward a dual purpose, milk-beef type:

A mature bull on a rainy day on an improved Sinú pasture. Montería, Colombia, 1975.

A dry cow, with horns that are typical of the breed but are rarely seen in the station herd since females are routinely dehorned. Montería, Colombia, 1975.

The station herd is being maintained in an effort to per-petuate the breed so that it may be used in crossing with the American Brahman. Twenty years ago an attempt was made to establish a breed society to maintain standards and initiate registration, but the idea was soon abandoned. The quick re-turns obtained from crossing the Zebu on the Sinú cattle were more attractive to most growers than the perpetuation of the breed. Unless the Colombian stockman can see the advantage in maintaining a base of pure Costeño con Cuernos cows to produce the first cross from Zebu bulls, the breed will probably disappear.

Romo Sinuano The polled Criollo, known as the Romo Sinuano (Polled from the Sinú), was developed in the Sinú region from the same base of old Spanish cattle as the Costeño con Cuernos. The only herd of any size today is the one at the Turipana Experi-ment Station near Montería where the Costeño con Cuernos herd, previously discussed, is also located. Again, the objective is to develop a base stock from which females can be obtained to produce a first cross Zebu-Criollo calf.

In all characteristics except horns, the Romo Sinuano close-ly resembles the Costeño con Cuernos. The Romo is a some-what larger animal and shows a slightly better beef conforma-tion, the Costeño possibly tending more toward a dairy type. The color of the Romo seems to be a slightly darker red, shad-ing less to the Jersey-tan. First developed as a draft animal, recently selection has been for a beef-type conformation.

There are two versions of how the polled feature was acquired: one, by the introduction of Angus or Red Poll bulls into an old breeding program; the other, by selective breeding from a polled mutation. There evidently is no record to sub-stantiate either theory, although there is definite evidence that the polled gene could have been introduced from either an Angus or Red Poll bull.

There is record of an Angus bull being taken to Antioquia in 1886, and four years later, one of his sons (presumably an Angus-Criollo cross) was moved to the Sinú by a cattleman named Durango. With this start, Durango is reported to have bred up a sizable polled herd. Another Sinú breeder, one Deriex, acquired polled bulls from Durango in 1914 and also developed a polled herd. Later, Deriex acquired bulls that were the offspring of Red Poll sires from Jamaica, and these were introduced into his herd. Around 1923 Deriex purchased a Red Poll bull directly from Jamaica to be used in his polled herd. There is also record of a number of Angus bulls being taken to the vicinity of Montería in 1923, but nothing is said about whether they entered into any breeding program. (Isaza, *Monografia*)

This fragmentary genealogy does not bridge the gap of the continuation of the polled descendants of the Angus and Red Poll bulls. It does indicate that a substantial number of polled animals carrying Angus or Red Poll genes, or both, existed in the Sinú 50 to 75 years ago. In support of the polled breed theory, it is known that many breeds of strongly horned cattle have produced polled mutations. This has occurred in the Spanish cattle. The herd of Chino-Santandereano at Bucaramanga has three naturally polled cows among 200 pure Criollos. A polled Criollo certainly could have cropped up and fostered a polled population. The absence of a written record of such a happening is not surprising. Historical records on Colombian cattle are few and far between.

An analogy applicable to the origin of the Romo Sinuano might be drawn from the evolution of the polled Shorthorn and polled Hereford in the United States. Around 1890 typical Shorthorn cattle without horns were developed by crossing with polled breeds. Ten years later a polled strain of Shorthorns was developed from a mutation that had purebred horned parents.

The Hereford breeders followed the same pattern—first, a polled population developed from crossing with a polled breed; next, a polled population built from Hereford mutations. Later, both the Shorthorn and Hereford breed societies attempted to bar from registration cattle that did not stem from the mutations of straight bred parents. A similar sequence of

The Romo Sinuano has nearly disappeared. The herd at the Granja Turipano station is an attempt to keep the breed alive. The Romo Sinuano is the foremost beef breed that was developed in Colombia and is the only polled Criollo breed:

This six-year-old bull weighed 1,485 pounds. Montería, Colombia, 1975.

A five-year-old cow that had recently freshened. Montería, Colombia, 1975.

developments could well have happened to the Romo Sinuano. There was no breed society to impose any barriers as to the origin of the polled cattle. The Romo Sinuano as seen today could have derived from the descendants of Angus or Red Poll bulls, or from a polled mutation of a Criollo, or from a combination of both.

Blanco Orejinegro   The most distinctive descendant of the Spanish cattle in the Western Hemisphere is the Blanco Orejinegro (white with black ears). This breed is usually abbreviated to BON (and Blanco Orejimono—BOM). It was segregated along the slopes of the western cordillera that rises from the Cauca Valley. This region became the major coffee-growing area of Colombia, stretching along the Cauca River from Antioquia in the north to Popayán in the south. It is bounded by the Cauca River on the east and by high mountains on the west. The BON is the traditional cow of the coffee planter, both a draft and milk animal. Now raised for beef rather than draft, the BON is still the farmer's milk cow. The environment is as ideal for cattle as can be found in the tropics, a warm but moderate climate, with good grasses.

In conformation, the BON is typically a dual-purpose, milk-beef breed. There is remarkable uniformity in the color pattern—white body, occasionally a few black spots around the neck and shoulders, and black ears. (On the BOM, the markings are red). The thin tail, high tail stock, thin dewlap of the Criollo are typical. Cows weigh 1,000 to 1,100 pounds, the bulls 1,400 to 1,600 pounds.

Twenty-five years ago the experiment station at San José del Nus on the Nus River undertook to introduce the Jersey influence into the BON to improve milk production. This experimental crossing was later abandoned and does not appear to have had any appreciable effect on the BON as seen today.

There are nearly as many versions of the origin of the Blanco Orejinegro as there have been commentators on Colombian cattle. The white cattle of Italy, the English Park cattle, the Fjällko of Sweden, the Blacksided Trønder of Norway, the white cattle in the Marianas, the white cattle on the island of Mauritius in the South Indian Ocean—all these, as well as numerous possible crosses, have been advanced as possible ancestors of the Blanco Orejinegro.

Why these writers went so far afield to locate a hypothetical progenitor for the BON is hard to understand. The obvious answer is still to be seen in Spain where all Criollos originated. While there are references to the ancestors of the BON coming from Spain, the Berrenda breed appears to have been overlooked. A Berrenda cow, as seen in Andalusia today, could be placed in the BON herd at El Nus Experiment Station and would pass unnoticed except for the fact that she carried horns. (Nearly all El Nus cows are dehorned.)

The Berrenda is an old Spanish breed whose ancestry must go back beyond the time of Columbus. There is a long blank in the record from the sixteenth century, when the first Spanish cattle arrived in Colombia, and the early twentieth century when there is written record of the BON. How this gap was bridged can be only surmised. It is not nearly as wide, however, as the one created by the assumption that the BON ancestors came from Scandinavia or the Indian Ocean.

A lone descendant of Berrenda parentage arriving in Hispaniola in the early 1500's, then transplanted to the Cauca Valley, could have founded the BON breed. The farmer who had this hypothetical early BON might have been attracted by his black-eared, white animal and started selecting for black ears. If a few of his neighbors perhaps did likewise, you have the foundation of the breed.

Possibly a more plausible hypothesis for the genesis of the BON would be a shipment of Berrenda cattle to the Indies in the sixteenth century. An odd lot of Berrenda cattle could have been loaded at Córdoba, the apparent homeland of the breed, off-loaded at Santo Domingo, transshipped to Panama, and finally arrived in the Cauca Valley. Kept separate because the unusual white body and black ear-marking appealed to an early enthusiast, such a group could have been the origin of the BON.

It is known that special shipments of cattle were made   *155*

The cattle of the coffee grower, the Blanco Orejinegro, with the white body and black ears, is the most distinctive of all Criollo breeds in Colombia. A herd of 500 head is maintained at El Nus Experiment Station:

This thirty-month-old bull, grown out on pasture, was estimated to weigh 1,350 pounds. San José del Nus, Colombia, 1975.

The general practice is to dehorn all females as calves at El Nus. This mature cow is an excellent example of the Blanco Orejinegro except for the loss of her horns. San José del Nus, Colombia, 1975.

from Spain, both to the Indies and the mainland, in early colonial times. The shipment, for example, of "selected cows and young bulls" to Puerto Rico in 1541 must have been arranged by an early Spanish hidalgo who wanted a new kind of stock for his herd. There were also the 24 head of fighting cattle that went to Mexico in 1552.

A moderate stretch of the imagination could envisage Berrenda cattle constituting the consignment which the conqueror Belalcazar took to the port at Buenaventura in 1541 and thence to the Cauca Valley. Belalcazar was born in Córdoba, and the black-eared cattle of his homeland could well have held a strong appeal. He had recently returned from Spain where his position would have made possible an authorization for a shipment of cattle to his capital. Whether the conqueror made such an importation will probably never be known. The similarity of the BON and the Berrenda is fair evidence that at some time in the past the black-eared cattle did reach the New World.

Coexistent with the Blanco Orejinegro there has always been the Blanco Orejimono (white with red ears). The difference between the two is only a single gene variation, as in the red and black Angus. In all other characteristics, the BON and BOM are identical. Thus, what is related here regarding the BON applies equally to the BOM, which at one time was said to be the more numerous of the two types.

It should be noted that pure or nearly pure Criollo with black ears and a white, or white and black, body are seen in such widely separated locations as Bolivia, Ecuador, Nicaragua, and Honduras. The circumstances under which these black-eared cattle were seen indicate that their heritage traces directly to the old Spanish cattle. Also, the black ears are seen in the reconstructed herds of Longhorn cattle in both Texas and Oklahoma. Such a Longhorn may be 90 per cent pure Criollo, and a portion of her genes could trace back to the early Berrendas that were brought to the Americas. For illustrations, see pages 277 and 280.

In late 1960's there were said to have been 2,000,000 head
of Blanco Orejinegro cattle in the coffee-growing area from

Antioquia to Popayán. Upgrading to the American Brahman then spread like wildfire throughout the region until the BON population in 1975 is said to be fewer than 200,000. Even so, this is still the largest number of Criollo cattle of one type in existence, but they are rapidly decreasing as crossing with the Zebu continues. The Ministry of Agriculture has a herd of 500 cows at the Nus Experiment Station and is making a determined effort to perpetuate the breed. Its objective is to create a base herd to generate females to be crossed with the Zebu. On the success of this venture depends the future of the BON.

The cattle in the Colombian Llanos today are predominantly upgraded American Brahman or criollo cattle showing varying degrees of Brahman influence. In the past other Zebu breeds have been used, but their influence may now be considered minor. In a few back areas of the llanos there still remain small concentrations of Criollos, pure or nearly pure. Llanos cattle are usually referred to by the name of the locality where they originated. The Casanareno, a pure Criollo, is found along the Casanare River in the northeastern llanos where a few old fincas have yet to complete the upgrading process.

Casanareno

    The Casanareno is a small animal, cows averaging possibly 600 to 700 pounds. Bulls are only a little larger than the cows. The predominant color is a Jersey-tan with lighter extremities. There are some solid blacks and a few mottled black and whites. The horns on the cow are large, wide, and upspread, often with a lyre twist toward the ends, and are black tipped. Horns on the bull are shorter and thicker. Unmistakably identifying the Casanareno as a Criollo of the tropics are the long, thin tail with a short black switch, high tail stock, a thin, prominent dewlap, barrel-shaped body, long muscular face, and a short, fine hair coat.

    By modern standards of bovine productivity, the Casanareno ranks very low. On the other hand, no modern European breed or crossbreed could survive long enough in the llanos to perpetuate its kind. The specialized gene pool which

A surviving type of the llanos Criollo, the Casanareno, can be seen along the Casanare River in the eastern llanos. These representatives were in a herd that had been gathered for branding. The introduction of Zebu bulls had just been started:

A young bull, weighing no more than 650 pounds, cut out for slaughter. Eastern llanos, Colombia, 1975.

This cow, a typical llanos Criollo, probably does not weigh over 600 pounds. Eastern llanos, Colombia, 1975.

The Lucerna is a new dairy breed built on a Criollo base, incorporating, first, Holstein-Friesian and, then, Milking Shorthorn influence. These representatives were in the foundation herd at the Lucerna Hacienda in the Cauca Valley near Cali, Colombia:

Lucerna bulls show the Shorthorn influence in the better rounded rump but retain many Criollo characteristics. This is a typical representative of the new breed. Cauca Valley, Colombia, 1975.

162

A high-producing cow (horns blunted) shows the Shorthorn influence. Her record is 7,000 pounds in one lactation. Cauca Valley, Colombia, 1975.

has been segregated in the Casanareno may be of no commercial use in cattle breeding today, but it could be of inestimable value in the not-too-distant future, when better use may have to be made of pastures as poor as the llanos.

The Casanareno is the end product of four centuries of natural selection. A hot climate prevails in the llanos. There are wide flooded areas during the May to November rainy season; sun-baked lands, dry water holes, and sunburnt grasses mark the remainder of the year. Continuous hatches of biting insects, including ticks, make life unbearable for any bovine other than the climatically adapted Criollo and Zebu. The soils are of low fertility although the land greens up quickly when the rains come. The coarse grasses have a low nutritive value and lack essential minerals. However, the low flooded areas furnish fair grazing during the dry season as do the higher lands that are not flooded in the rainy season. Such conditions produced the Casanareno, an extremely hardy bovine type, highly resistant to insects and heat, extremely fertile, and tolerant of mineral and nutrient deficiencies.

Lucerna   A new breed incorporating the Criollo, Holstein-Friesian, and Shorthorn breeds has been developed at the Hacienda Lucerna, north of Cali in the Cauca Valley. The Vallecaucano was first crossed with Holstein-Friesian bulls. After a number of years the mixed female progeny was bred to Milking Shorthorn bulls. The breed composition is not definitely known, but is estimated as 30 per cent Criollo, 40 per cent Holstein-Friesian, 30 per cent Milking Shorthorn.

Color of the Lucerna at its present state of development is predominantly a solid, or nearly solid, cherry red. In the 2,000-head cow herd, there are several black animals or black with some white, particularly on the underline. Conformation is of the dual-purpose type. Horns on the bull are thick, short, and outthrust, similar to the Shorthorn; on the cow the horns are thin, quite short, and upthrust. Mature cows weigh 1,000 to 1,100 pounds, and bulls average 1,600 pounds. Milk production with little grain feeding is said to average 4,400 pounds for a 305-day lactation with some individuals yielding a maximum of 7,700 pounds.

Selection in the Lucerna has been for milk production, tempered by some attention to udder development and a milk-type conformation. Considerable Shorthorn influence appears to have been retained in the well-fleshed three-way cross. The Lucerna is one of the few efforts of any moment to combine northern European breeds with the Criollo to form a new breed.

Mexico

The farmers and cattlemen in nineteenth century Mexico do not appear to have had the interest in selecting and maintaining distinct cattle types as was done in Colombia. The Criollo cattle became adapted to widely varied environments. Husbandry practices did not lead to a marked differentiation into what could be called breeds although a distinct dairy type Criollo did emerge in southern Mexico.

The first Zebu to enter Mexico arrived from Brazil in 1884. The Mexican cattlemen soon learned, as had the Jamaican breeders, the art of producing a larger draft animal by using humped bulls on the Criollo cows. Breeding to Zebu bulls gradually spread along the coastal areas. The Nellore, Gyr, and Indo-Brazil were the favored breeds until the time of World War II. After the war, the American Brahman, developed from Zebu breeds brought to the United States mostly from Mexico, caught the attention of the Mexican cattleman. By this time the Criollo on the coastal plains of Mexico was disappearing, and the Brahman bulls were used in herds of criollos that traced to Zebu sires of other types. Purebred herds of the Indo-Brazil and Nellore are still maintained, however, and these breeds continue to be used.

The Criollo herds on the high plateau region, extending southward from the Rio Grande to the broken country north

Chinampo cow in the holding yard at La Paz, Baja California. The conformation of this Criollo type bears a marked resemblance to the Florida Scrub. La Paz, Mexico, 1968.

of Mexico City, were graded up by the use of Hereford, Shorthorn, and later, Angus bulls, with the Hereford predominating. Also the Charolais, imported from France, became established in Mexico following their introduction in 1930. The limited number of owners were mainly interested in upgrading their herds to straight Charolais. Founded on a base of Criollo cows, the Mexican Charolais became well adapted to the hot climate. In number, however, the French cattle were not a major factor in breeding out the Criollo.

Around the large cities, the northern European dairy breeds took over; the Holstein-Friesian and Brown Swiss were the most popular in the organized dairies. Bulls of these breeds were also used by small farmers on their Criollo cows whenever such service could be obtained.

By the late 1960's the Criollo had disappeared from Mexico except for small isolated groups in the hinterlands and a few survivors on the southern end of Baja California where they are known as Chinampo cattle. One organized herd of Milking Criollo, to be discussed later, was established in 1965.

A few scattered groups of the Chinampo, held by small owners, have survived in the desert region of Baja California with very few ancestors of foreign breeds in their unrecorded pedigrees. Over a period of nearly three centuries, natural selection produced a small, extremely hardy beast, a browser on desert plants when the scant grass disappears, with the ability to exist on cactus without the thorns being burnt off. In recent years American Brahman and Charolais bulls have been used by the few large ranches on the lower peninsula that are now under irrigation. Apparently there has been no other influence of outside breeds.

The Chinampo as seen today does not display the Jersey-tan color of most Criollos. The common pattern is white with black or red markings, or nearly all white. An occasional individual is all black. In conformation the Chinampo bears a marked resemblance to the Florida Scrub cattle. Cows weigh 600 to 700 pounds in fair flesh, a condition in which they are rarely seen. Bulls outweigh the cows by possibly 100 pounds.

Chinampo bulls are being eliminated as rapidly as the small farmer can obtain a crossbred Zebu replacement. The type will probably not survive another decade, and with its disappearance will go one of the hardiest bovine types which natural selection has ever produced—an irreplaceable ruminant which could someday have made a significant contribution in an overcrowded world.

In remote areas of the states of Oaxaca and Guerrero a few pure Criollo cattle could still be found in the late 1960's.   *167*

These were the property of small growers, usually engaged in a village dairy operation. In appearance these descendants of the old Spanish cattle closely resemble the Milking Criollo of Nicaragua. Their progenitors were probably chosen because they had produced a better calf than their herdmates. Some local situation—isolation from other cattle or the lack of a large ranch in the vicinity with imported Zebu bulls—accounted for these forgotten islands of Criollo. Better roads and the growing improvement in the rural economy of Mexico will effect their disappearance in a few more cattle generations.

One organized herd of Milking Criollo existed in Mexico in 1975, the result of ten years of effort by Dr. Jorge de Alba to perpetuate a nucleus of this remarkable animal in his country. At his instigation the Asociación Mexicana de Producción Animal, a private organization devoted to improvement in animal production, had sponsored a program in 1965 for the establishment of a herd of Milking Criollo. The herd at Turrialba, Costa Rica had been founded by Dr. de Alba in 1950. (See page 179.)

The initial move was the shipment of 18 cows by air from the Rivas area of Nicaragua to Veracruz. These cows, bearing the clover leaf brand on the left hip and "R" on the right, were obtained from the widow of Don Joaquin Reyna. (See page 178.)

Two sires, the sons of the highest-producing cows at Turrialba were a gift from the Instituto Interamericano de Ciencias Agricolas (IICA). This small herd was placed on the Rancho Rio Florida in the San Luis Potosí area under a maintenance agreement by which the ranch owner retained the bull calves and the Association owned the heifers. The herd, with the increase, passed to several ranchers in the area under the same arrangement for maintenance, and in 1972, was finally placed on Rancho El Apuro on the coastal plain forty-five miles north of Tampico. This was an undeveloped property of 1,000 acres that had been acquired by the Association as a training center for graduate students in the fundamentals of animal production research.

A Criollo cow, raised in the Mexican state of Guerrero, was brought to Rancho El Apuro in 1972. She is typical in all Criollo characteristics, with horns bearing a marked resemblance to those of the Retinto in Spain. Tampico, Mexico, 1976.

From the pockets of Criollo animals in the states of Oaxaca and Guerrero, 35 of the best representatives were added to the herd at El Apuro. The only bulls used, however, were those tracing to the Reyna or Turrialba herds. In 1972 natural service in breeding was discontinued, and all cows bred artificially to selected Turrialba bulls which were the sons of cows with production records in the range of 6,500 pounds per lactation.

The herd is being maintained as pure Milking Criollo and selected for milking capability by production testing. The cows are beautiful representatives of this tropical dairy animal. Many show the reddish cast that traces to the Reyna breeding. In this remarkably uniform group, the true Criollo characteristics are in evidence: the short, fine hair coat, thin tail with a short   *169*

These descendants of the Reyna cattle in Nicaragua were from the herd of Milking Criollo at the Rancho El Apuro, north of Tampico, Mexico:

The bull, Pálido, had recently been sold to a breeder on the Pánuco River. He was photographed at the close of the breeding season. Pánuco, Mexico, 1976.

The cow, Huerfana, seven years old, was seen on good Coastcross 1 pasture in January. Tampico, Mexico, 1976.

brush, wrinkles over the eyes and on the neck, a barrel-shaped body with parallel top and bottom lines, and wide horns up-thrust at the ends.

Experimental work is underway at El Apuro looking toward the development of a crossbred dairy animal combining the milking capability of the Brown Swiss breed in one line and Jersey in another, and in both lines the natural resistance of the Criollo to tropical diseases and heat tolerance. The Milking Criollo herd, however, is to be maintained pure as the source of females to be crossbred and bulls to be used locally on females that have been graded up to Jersey or Brown Swiss types.

Throughout Latin America perpetuation of the Criollo is generally in the hands of government-sponsored activities— experiment stations, a university herd in Colombia, the state farms in Cuba. Three exceptions are the Apuro herd, that of the local cattleman's association out of Bucaramanga in Colombia, and the Criollo herd on the Espiritu ranch in the Beni of Bolivia. These are the only privately financed attempts to preserve the line of descent of the old Spanish cattle that are known to me. Continuity of government programs involved in animal production is subject to many political hazards. These three endeavors to continue the Criollo line could well be the determining factor in its preservation.

Central America

The Criollo of Central America remained uncontaminated with any other cattle types for nearly four centuries after they were first brought to the shores of Lake Nicaragua. The introduction of the Zebu into Central America began early in the twentieth century with the usual motive of obtaining a better draft animal. About the same time representatives of the European dairy breeds were brought into the areas around the larger cities.

172

The Jersey and Brown Swiss were the most popular originally although there were several local concentrations of Guernsey and Red Poll. The Holstein-Friesian came in later.

The breeding out of the pure Criollo proceeded at varying paces in the several small countries and increased rapidly after World War II. Central America then began to enjoy a favorable export market for beef to the United States. This led to increased importations of American Brahman and Santa Gertrudis bulls. Purebred herds were established for the production of bulls to be used in the criollo herds.

By the late 1960's the cattle population was largely criollo showing varying degrees of Zebu influence. In all of the countries (Panama not included) representatives of essentially pure Criollo could still be found in the hinterlands or occasionally be seen in the holding yard of an abattoir or at a cattle market. Costa Rica had in addition to a few Criollos in the hills a small herd that had been selected for milk production. This group will be discussed later.

The unselected Criollo cattle now existing in Central America are found in pockets in the mountainous regions where they have been more or less isolated from other types for many generations. Even little El Salvador, with the highest population density in Central America, can still display an occasional true Criollo. These animals are fair representatives of the kind of cattle that evolved in the Central American environment from the original Spanish introductions.

The size of the Central American Criollo varies widely. In dry areas, where an owner's few head run loose in the hills, the cows weigh 500 to 600 pounds. The local dairyman, selling a few liters of milk in his village, sees that his cow, which may weigh over 800 pounds, is on the best grass available. Color is also varied, but the same patterns appear throughout all the small Central American countries where the Criollo can still be found. The Jersey-tan predominates, then solid black, and black-and-white mottled. There are few reds. In Guatemala and Honduras, the odd cow in a herd will have a nearly white hair coat with some black spots. Occasionally one with a white

An upgraded American Brahman-Criollo cow herd collected for pregnancy testing in the Rivas area, Nicaragua, 1969.

A village dairyman's Criollo cow, offered for sale with calf by side, in the San Salvador cattle market. Since the presence of the calf is considered necessary for the cow to let down her milk, it sells with the cow. El Salvador, 1969.

These two Criollo cows probably descended from lines of progenitors separated for countless generations. While their respective herds were not more than 150 miles apart, the rough terrain made any communication practically impossible until very recent times. The tie between the two animals undoubtedly goes back to the day when the ancestors of the present owners moved out on the land:

The Criollo cow, with slightly blunted horns, of a farmer in western Guatemala, 1969.

A young cow of a small owner in eastern Honduras displays the characteristics of the true Jersey-tan Criollo, 1969.

body and solid black ears is seen, the markings of the Blanco Orejinegro of Colombia.

In Costa Rica and Nicaragua a milking-type Criollo was developed in the coastal plains along the southwestern shore of Lake Nicaragua. The area became noted for its hard cheese, which item of commerce was probably the reason the farmers there culled their herds to obtain a line of good milking cows. The results of the selection practices followed by these dairymen paralleled the accomplishments of those in the Rio Limón area of Venezuela and the Cauca and Lebrija valleys and the Sinú region of Colombia. The Limonero around Lake Maracaibo, the Vallecaucano, Chino-Santandereano and Costeño con Cuernos of Colombia, as seen today, show a marked similarity to the remaining descendants of the Lake Nicaragua Milking Criollo. The Criollo of Cuba, which has also undergone selection for milking ability, must be included in this family even though bearing a trace of Zebu influence.

Central America is the one area on the continent where pure descendants of the old Spanish cattle which have been subjected to selective breeding can be directly compared with those which have come down through natural selection. In the mountainous areas of eastern Honduras and western Guatemala life goes on much as it has for the past few centuries. The farmer who has risen a little above the subsistence level has an ox team and his milk cow. The influence of a Zebu or Brown Swiss bull is seen in the cattle in the hills, and is increasing, but in the late 1960's the pure Criollo cow could still be seen. Not much more than half the size of her sister that came from a dairy herd around Lake Nicaragua, her wrinkled forehead, barrel shaped body, and thin, Jersey-tan hair coat brands her as a true relative. Here, again, the descent from the same genetic base cannot be questioned.

During the early years of the twentieth century the Criollo around Lake Nicaragua appear to have been concentrated in the Rivas area of Nicaragua. Two well-known herds were in existence there around 1910—one at San Rafael, a home for the aged, and another that had been developed by Don Joaquin

Reyna, an early selective breeder interested in high milk production for the manufacture of cheese. The best dairy-type cows available in the vicinity were purchased by Don Reyna and taken to his ranch on the Bahia de la Flor, near the Costa Rican border. One of the selection procedures with which Don Joaquin is credited was to cull any cow which did not yield a fair quantity of milk in the pail after her calf had sucked. He was also partial to a red color, and descendants of the Reyna herd today still have a distinctive reddish tinge instead of the more typical Jersey-tan of the Criollo. Small, thin horns were also a selection criteria.

In 1950 a program for further development of the Rivas Criollo was begun at the Instituto Interamericano de Ciencias Agricolas (IICA) at Turrialba, Costa Rica. Dr. Jorge de Alba, an animal scientist at the Institute, held the concept of increasing the milking capability of this climatically adapted Criollo by modern selection practices. He chose the first cows for a foundation herd at Turrialba from San Rafael; later additions were made from the Reyna herd. Outstanding individuals were also obtained from other places in Central America where Dr. de Alba could locate animals exhibiting true Milking Criollo characteristics.

The Turrialba herd was eventually increased to 300 producing cows. The average milk production reached 4,400 pounds of 4.6 per cent butterfat content, with individuals yielding 9,400 pounds in a 305-day lactation. Later, the herd was reduced to 50 of the highest producing cows, and the bulls used for breeding were sons of the top cows. Because of lack of facilities for progeny testing, the bulls were discarded at two years of age before their ability to transmit milking capability could be determined.

The Turrialba cow is a beautiful example of what modern selection procedures can develop from the Criollo which has undergone many generations of the farmer's effort to increase milk production by observing the quantity of milk left in the pail. The fine, short hair is a reddish Jersey-tan, sometimes approaching the cherry red of the Retinto. Other typical Criollo   *179*

The Milking Criollo were developed at the Instituto Interamericano at Turrialba. This herd closely resembles the Limonero cattle of Venezuela:

A thirty-month-old bull, partly dehorned, used as a sire. Turrialba, Cost Rica, 1969.

A high-producing mature cow that had been dehorned. Turrialba, Costa Rica, 1969.

A representative cow in a Criollo herd in eastern Honduras, 1969. The herd was under selection for milk production.

features are also displayed—thin dewlap, muscular face with wrinkles over the eyes, thin tail with a short brush. Cows in good milking condition weigh up to 1,100 pounds, mature bulls 1,500 pounds.

The experiment station at Turrialba for years has had a herd of Romo Sinuano, the result of an unusual movement of Criollo cattle. Before the day when the importation of cattle to the United States from Latin American countries was pro-hibited by the United States Department of Agriculture quarantine regulations, a small group of Romo Sinuano cattle in Colombia was brought to an experiment station of the University of North Carolina. When Dr. de Alba learned that the project in which these cattle were involved was to be terminated, he managed to obtain four bulls which were brought to Turrialba in 1957. By breeding these to Criollo cows a line of polled cattle that was predominantly Romo Sinuano was established. An occasional naturally polled cow which showed up among the Milking Criollo herd was transferred to the Romo herd. This reconstructed herd of Romo Sinuano, now numbering 50 head, is being maintained as a Criollo beef type.

In Honduras, a few growers have made some selection for a dairy-type Criollo. In the mountainous country east of Tegucigalpa an occasional herd can be seen that is being developed for better milk production. The best producing cows are Jersey-tan or mottled black and white. They will average 950 pounds in weight with the top cows yielding 3,500 pounds of milk per lactation, with no grain fed.

Efforts to maintain a line of pure Criollo cattle of the kind mentioned in Costa Rica and Honduras are highly exceptional in Central America. Such efforts are the work of a very few dedicated individuals.

United States

The Native American cattle followed the western movement of population and had reached Louisiana in the early years of the nineteenth century. There the first significant interbreeding of Spanish cattle with the Native American occurred. This movement, however, did not extend into East Texas until after Mexico gained her independence in 1821. Many Anglo-Saxon settlers then swarmed onto the Texas plains bringing their cattle with them. Again the Native American and the Spanish cattle met. Finally in the 1850's the Native American reached

The vanishing Florida Scrub could still be found in the early 1970's. These representatives, from the herd of an old-time cattleman, are still held by the James Durrance Estate:

A Jersey-tan colored cow weighing 550 pounds. Bassinger, Florida, 1971.

Another cow, almost red in color, on good Pangola grass pasture. Bassinger, Florida, 1971.

the California range and the final contact with the Spanish cattle was made. (For Native American Cattle see page 91.)

The British beef breeds, following on the heels of the Native American, were responsible for the disappearance of the Spanish cattle from the United States. After the Civil War, Shorthorn and Hereford bulls were put on the herds in the Southwest as rapidly as they could be obtained. The upgrading pattern varied in different parts of the country but continued relentlessly until the Spanish type all but disappeared.

Florida   In a narrow belt along the Georgia-Florida boundary there had been only minor mixing of Native American and Spanish cattle by 1800. The northern cattle could not withstand the Florida environment, and the farmers north of the border had no use for the Florida Scrub. These hardy beasts, which had completely adjusted to their harsh surroundings over a period of two centuries, were now to enjoy another century without serious contamination by other cattle.

The Florida Scrub is the product of natural selection and uncontrolled breeding in an environment characterized by a hot humid climate and forage low in nutrients, deficient in proteins and minerals, and often inadequate even in quantity. The cow is a small animal weighing 450 pounds on palmetto land, 650 on prairie ranges. Bulls are 100 pounds heavier. The horns are of good size but not as large as those on most descendants of the old Spanish cattle. The middle is barrel shaped, and the legs are long and thin. The tail is long with a heavy brush. The color is widely varied—solid red, black, or Jersey-tan, mixed red and white or black and white, a few brindles, and sometimes there is a white topline. The Scrub cow lacks some features typical of the Spanish cattle—the large horns already mentioned, the thin tail, prominent dewlap, and the predominance of the Jersey-tan color. Deficiencies in the diet for many generations account for the small size.

The Florida cattleman, well aware of the inferior quality of his stock, had made many attempts before 1900 to introduce the British breeds into his herd. At first, such efforts met with complete failure. The imported bulls could not survive the onslaught of ticks, biting insects, and heat stress. There is an account of a carload of Hereford bulls from Texas being turned loose on the Alachua range in central Florida. At the next roundup it was found that only three had survived.

The Zebu made its first appearance in Florida in 1880 but was not used to any great extent for another fifty years. Also, after 1900 a few purebred herds of the Hereford, Shorthorn, and Devon breeds were established and maintained under special care. It was the control of the cattle tick in 1931 that paved the way for substantial progress in upgrading the Florida herds. To acclimate and maintain the northern bulls in breeding condition, however, still required exceptionally good management—dipping, nutritious forage, minerals, and immunization against diseases.

The Florida cattleman, accustomed to working with the Scrub that required no pampering, was slow to change his operation. The hardiness of the Scrub was an asset the old-time cattleman treasured. Eventually, the American Brahman came to have the largest influence in breeding out the Spanish cattle. Requiring no more care than the Scrub cow, the Brahman bulls gradually came into general use in the beef herds. By 1970 only a few small isolated herds reminiscent of the old cattle could be found. These were usually without Scrub bulls to perpetuate the type. A movement to segregate and preserve a herd of this remnant of the Spanish cattle failed—the gene pool of the Florida Scrub was doomed.

The number of Native American cattle to reach Louisiana had been insignificant before 1803, when the United States acquired that vast territory from France. And the French Canadian cattle that the Acadians had brought to the area had little influence on the existing cattle population. Thus the Louisiana cattle in the early years of the nineteenth century were still essentially pure descendants of the old Spanish cattle. One traveler of that

Louisiana

187

day describes them as having ". . . extraordinary fine horns about 2-1/2 feet long," and of the "usual red brown color"; another, as "high, clean limbed, and elegant in appearance." (Post, *Cattle Industry of Louisiana*)

Sizable movements of Native American cattle flowed into Louisiana in the first quarter of the nineteenth century. These came by boat down the Mississippi River and overland across the borders with Mississippi and Arkansas. The large cattle holdings of the preceding century were disappearing as the grasslands that supported the herds were put into cotton, rice, and cane. Farmers saw that the Native American when crossed with their Spanish cattle gave a better draft animal and a more productive cow. Before the middle of the nineteenth century, the Spanish cow had been practically bred out, and in her place a nondescript Native American appeared, showing little evidence of her Spanish heritage.

California   Native American cattle were brought into California in the early 1850's in considerable numbers. The gold rush had created an inordinate demand for beef, and cattle were trailed in from wherever they could be obtained. Rail shipments were not practical until 1870. Most came from the Midwest, but New Mexico and Texas also moved large numbers of cattle to California at this time. Before the inrush of gold seekers, the only outlet for California cattle was in hides and tallow. When the market for beef opened, the advantages of a larger, better-fleshed animal were soon recognized.

Some of the first Native American cattle brought to California went into breeding herds and started the upgrading of Spanish cattle. After the Civil War the craze for the English purebred reached the West. Grade Shorthorn and Hereford bulls, as well as bulls of some dairy breeds, were brought in. Crossbred cattle soon predominated in the herds of the large ranches, and before the end of the century the Spanish cow had disappeared, replaced mostly by the Hereford and, to a lesser extent, by dairy breeds.

The Hawaiian Wild Cattle, shown here and on the following page, are now probably extinct, but in the early 1970's rare individuals could still be seen on the slopes of Mauna Kea on the island of Hawaii.

A bull in a remote mountain area of dense vegetation. The bushy tail is not typical. Hawaii Island, 1972.

For years the Hawaiian Wild Cattle, descendants of the Spanish cattle which had arrived on the island of Hawaii from California, ran free on the slopes of Mauna Kea. They were slaughtered by the landowners largely as a means of ridding their lands of a nuisance. The Hawaiian cowboy, as experienced as any vaquero in Mexico, would rope an animal and leave it tied to a tree. Two or three days without water or forage subdued the animal to the point where it could be driven to

Hawaii

A cow, in heavy timber, shows conformation similar to that of the Chinampo in Baja California. Hawaii Island, 1972.

market. In the early years, however, the demand was mostly for hides and tallow.

The Shorthorn was introduced as a dairy animal after the middle of the nineteenth century. Somewhat later other beef breeds were imported. At first, efforts were made to keep these imports separated from the wild cattle, though there was some upgrading by replacing the bulls of the wild herds with pure-bred sires. Finally, the wild cattle were fenced off the grazing lands of the domesticated herds.

The Hawaiian wild cow was never redomesticated. She was eliminated much as a predatory animal, when farming and ranching moved farther up the mountains. In the early 1970's a few survivors were still to be seen in remote mountain areas. These appeared to have suffered very little contamination from other breeds. The remaining Hawaiian Wild Cattle were usually black or a light tan, although other mixed-color patterns appear. Horns are small. The cows weigh up to 700 pounds, and the bulls are not much heavier.

Before 1821, the year Mexico gained independence from Spain, there was no significant movement of cattle from the United States to the area that eventually became Texas. Stephen Austin, son of Moses, who followed through on his father's scheme to lead a group of settlers into New Spain, arrived there only after Mexico had gained independence. When the Austin party trekked into Texas, the hazards from Indian attacks and troubles along the trail precluded their bringing many cattle. **Texas**

When Texas became Mexican territory, emigrants from the southern states began moving into East Texas, bringing their ox teams and their milk cows. After Texas gained independence from Mexico in 1836, immigration increased rapidly, moving as far west as the plains area below San Antonio. A larger wave of settlers followed when Texas joined the Union in 1845. Bankers, lawyers, tradesmen, mechanics—men from all walks of life—stampeded to Texas for the next two decades. Many were farmers from the southern states who brought along breeding stock, occasionally herds of as many as one hundred head. These were all Native American cattle.

No estimate is available of the number of Native American cattle which entered Texas during the first 25 years of American immigration. There were said to be 30,000 United States citizens in Mexican Texas when the Republic was established. This alone would indicate the presence of a sizable number of Native American livestock. When Texas became a state nine years later, the population was estimated to be 125,000 to 150,000. The influx from the States accounted for most of this increase. Native American cattle holdings must then have reached into the tens of thousands.

191

These cattle grazed at first on the outer fringes of the much larger population of Spanish cattle out on the open range. Interbreeding was limited. As the new landowners moved westward, the area of contact widened and more interbreeding occurred. By the middle of the century, the Spanish cattle could be counted in the millions; the Native American might by then have reached into a few hundred thousand. These were closely confined as many of the animals were in daily use for milk or draft. The Spanish cattle were still on the open range. Under such conditions the degree of mixing between the two types cannot even be approximated. Although small, it did occur.

Evidence of this interbreeding was seen in the wide range of color patterns on the "Longhorns" as they came out of Texas. Their varied colors have often been described—red, black, black-and-white, red-and-white, and various shades of brindle animals. Old photographs confirm patched color patterns and the brindles.

More concrete evidence of a degree of mixing between the Spanish cattle and Native Americans is given in the blood-typing work of Kidd. *(Breed Relationships)* This shows that the best representatives of the Longhorn extant today—those segregated and selected for Longhorn characteristics for fifty years—show a basic Spanish genetic development, influenced to a significant degree by northern European cattle types.

When the large drives of Longhorns to the northern railheads began in the late 1860's, the population from which they came varied from pure Spanish cattle to the progeny of the indiscriminate crosses which continued to carry a predominant Spanish influence. No distinction was made in the conglomerate drives that moved north—all were simply "Texas Longhorns." Basically, they were Criollo cattle, modified by natural selection during 350 years in North America, on which a small degree of crossing with northern European cattle had been imposed for a few decades.

The long drives in the 1870's out of Texas initially consisted of steers and bulls with sometimes a few old cows, all destined for Kansas City and the Chicago packing plants. This left a surplus of females on the range, and drives of cows to establish breeding herds on the open grasslands from Kansas west to the Rocky Mountains soon followed. By this time, Shorthorn bulls were becoming available from eastern breeders. The newly established cattleman of the western plains soon learned that he got a heavier steer by using these bulls, and the demand for them soon outran the supply.

Thus the decline of the Longhorn was accelerated. The process extended through the Longhorns in Texas and up into recently established herds in Montana and Canada. By 1880 the Hereford population in the east had been increased to the point where bulls of this breed became more available to the western cattleman and began to breed out the Shorthorn-Longhorn cross. The first quarter of the twentieth century saw the upgraded Hereford take over from Texas to Canada.

The Longhorn had now disappeared except for a few small isolated herds maintained by nostalgic oldtimers. Before its extinction, the famed animal was saved by action of the United States Department of the Interior. In 1927 a herd of 21 Longhorns was segregated at the Wichita Mountains Wildlife Refuge near Cache, Oklahoma, selected as the best representatives that could be found during two years of scouting through southern Oklahoma and southwest Texas. Later, another herd was founded at the Wildlife Refuge at Fort Niobrara, Nebraska from progeny of the Oklahoma herd. Then the state of Texas established herds in some of the state parks. Recently, Longhorn fanciers have started small private herds, and a breed society has been formed.

These modern Longhorn herds are very fair representatives of the old Texas cattle, but, on good pasture and under controlled breeding for fifty years, they have undoubtedly increased both in body and horn size. The larger horn probably results from specific selection for that feature. The most prized individuals in the present herds often have horns that surpass many of those seen in old photographs of the northern drives.

The Texas Longhorn appears to have been perpetuated

The modern Longhorn has now been selectively bred for a distinct type for fifty years. The Longhorn was saved from extinction by the establishment of a small breeding herd of 23 head in 1927 at the Wichita Mountains Wild Life Refuge, Cache, Oklahoma:

A Longhorn bull on the Jack Phillips Ranch, South Texas, 1971.

A Longhorn cow in the herd at the Wichita Mountains Wild Life Refuge, 1971.

in the descendants of the small group selected by the Interior Department back in 1927. It must be remembered, however, that the true Longhorn was a hybrid that carried a high percentage of Spanish genes combined with a small influence of the Native American. Also, having been selected later, the present Longhorn may have been contaminated a little more by the northern European breeds than was its predecessor.

The Indies

Jamaica
The English plantation owners of mid-nineteenth century Jamaica were aware of the capability of the Zebu as a draft animal from the experience of their compatriots in India. Slave labor had been abolished in 1832, and the prospering sugar economy needed energy. The Criollo (Creole in British territory) ox team was never the equivalent of the Zebu before the plough or cart. The Jamaican cane planters appear to have been among the first in the New World to see the advantage of cattle for plantation work and to power the cane mills. They began importing the humped cattle from India in the early 1860's in order to increase the draft power of the Criollo.

The English planter often had a personal preference for the breed of Zebu he worked, and the Hissar, Kankrej (Guzerat), Mysore, Ongole (Nellore), and Sahiwal were all brought in. The Zebu-Criollo cross from any of the Indian breeds proved a very acceptable draft animal. Here the fate of the Criollo on the hemisphere was sealed.

Early in the twentieth century English dairy and beef breeds were brought to Jamaica, particularly the Jersey, Red Poll, and Shorthorn. These also contributed to the breeding out of the Criollo. The tractor began to take the place of the draft ox in 1940. Interest then focused on the development of dairy and beef herds. After the middle of the century, all trace of the old Spanish cattle had disappeared from the island.

196

Puerto Rico, the next island after Cuba to receive Spanish cattle in the days of conquest, was the next to see them go. The Criollo dominated on the island until it joined the United States after the Spanish-American War. Special mention is made of a Durham (Shorthorn) bull that was imported about 1860. There were a few later importations of Shorthorn, Hereford, and Jersey cattle but these were not in sufficient numbers to have any significant effect on the island herd, which in 1899 numbered about 260,000. Pictures of Puerto Rican cattle of that day reveal animals of distinctly Criollo type. *(Report, Bureau of Animal Industry, 1901)* Crossing with the Zebu for a better draft animal had begun, but the Criollo still predominated in the large herds of the sugar plantations.

Puerto Rico

After World War II the industrial development in Puerto Rico proceeded rapidly and, again, the tractor replaced the draft ox. The demand for milk and beef soon exceeded the productive ability of the island herds, and importation of these products from the United States was necessary. Cane land was converted to pastures; the Criollo was replaced by the Holstein-Friesian in the milk barn and by the American Brahman for beef.

With variations in timing and breeds, the cattle cycle of Jamaica and Puerto Rico was repeated on the other islands of the Indies with the exception of Hispaniola and Cuba. The small islands of the Lesser Antilles were slower in eliminating the Criollo but followed the same breeding-out pattern. American Brahman beef herds and Holstein-Friesian dairies now dominate the cattle scene. Occasionally the Criollo cow of a small farmer could be seen grazing along the roadside on a few islands as late as 1970. The Criollo bulls, however, had gone to market, and the current generation appeared to be the last pure Criollo.

Lesser
Antilles

The husbandry of cattle on Hispaniola has varied widely between the Dominican Republic, which occupies the eastern side

Hispaniola

The true Criollo, descendants of the first cattle to reach the New World, continue to survive in small herds, particularly in the more remote areas of the Dominican Republic:

A young bull that had been brought to one of the Santa Domingo abattoirs. His owner was probably a small farmer in need of cash. Dominican Republic, 1970.

A well-kept Criollo cow, weighing about 850 pounds, in a small dairy herd west of Santo Domingo. Dominican Republic, 1970.

The Romana Red was developed by the Romana Central Corporation to obtain a better draft animal for plantation work. The Criollo was crossed with the Mysore and Nellore Zebu breeds of India to produce the new breed:

A ten-year-old Romana Red bull weighing 1,800 pounds. The hump is noticeably smaller than on the true Zebu as a result of Criollo influence. Dominican Republic, 1970.

200

An eight-year-old cow, weighing 1,100 pounds. The Romana Red cow has the reputation of being more fertile than the Zebu and also having a more tractable disposition. Dominican Republic, 1970.

of the island, and Haiti on the west. During the past century, there has been very little communication between the two countries.

In spite of its recurring political problems, the Dominican Republic displays a modern-type economy around Santo Domingo, the capital, which still carries the same name as when founded by Columbus' brother Bartholomew. Agriculture has reached the tractor stage on the large holdings. In the back country the small farmer carries on about as his forbears did at little more than a subsistence level.

Most of the cattle in the more advanced agricultural areas are grade Zebu. There has been some crossing with Charolais bulls and also with new breeds carrying a strong Zebu influence, such as the Brangus, Santa Gertrudis, and Charbray. The grade American Brahman, however, bred up from the Criollo, predominates in the large cattle holdings. Away from the more populated areas, the criollo, an indeterminate cross resulting from the use of American Brahman and northern European bulls, are the cattle of the country. Interspersed within the small cattle holdings, many representatives of the pure Criollo can still be seen. The Jersey-tan color predominates, but black-and-white patterns are more frequent than in many Criollo populations. A cow on good pasture weighs around 800 pounds, a bull 900 to 1,000 pounds. Conformation and other characteristics are typically Criollo.

Representative Criollos will probably continue to exist in small numbers in the back country for several generations. This will not be by design but because the isolated owner either lacks access to a bull of the dairy or Zebu breeds or does not bother to get his cow to one. There is no movement by government or any other organization to perpetuate the old Spanish strain. Eventually the Criollo appears certain to disappear from the Dominican Republic.

A new breed embodying both Criollo and Zebu genes, known locally as the Romana Red, should be noted. Before World War II the Central Romana Corporation, a large planta-

202

tion operator, imported a number of Criollo cattle from Puerto Rico. These were said to be of a "bright red color." The Criollo cows were bred to both Mysore and Nellore bulls in planned matings. The goal was to produce a better draft animal than had resulted from indiscriminate crossing of Zebu and Criollo cattle. Selection was also for a solid red color. A fixed cross, that proved to be an excellent draft animal, was developed. Other plantations made some use of the new Romana Red breed. When the tractor came in, the need for work oxen diminished. Breeding stock was then selected for a beef conformation, but the solid red color was retained.

The Romana Red in 1970 was a large animal for the tropics, cows weighing 1,100 pounds and bulls up to 1,800 pounds. The color was a solid cherry red. Some Zebu characteristics were obvious but had been considerably modified by the Criollo influence. Both cow and bull carried a distinct hump which was smaller than on a true Zebu. The long drooping ears of the Zebu were lacking although the ears were larger than on a Criollo. Horns were typical of a Zebu-Criollo cross, thicker and not as widespread as on the Criollo. The well-rounded rump lacked the characteristic slope seen on most Zebu cattle.

Throughout tropical Latin America there has been little interest in the development of new breeds by crossing the Criollo with other types. The Tipo-Carora, a Criollo-Brown Swiss cross, was started in Venezuela but, as has been seen, was soon bred out. The Lucerna, the Criollo-Holstein Friesian-Shorthorn cross, mentioned under *Colombia*, is being continued through the efforts of a single family. Crossing the Romana Red with Charolais was gaining favor in 1970 to obtain a better beef conformation, and it appeared that the breed would probably disappear.

Haiti

The low economic level in Haiti, isolation from her neighboring country on the island, and the large proportion of urban population probably account for a minor interest in cattle. The Criollo has not been crossed with other breeds to the extent that has occurred in most Caribbean countries. The Zebu was

The Criollo in Haiti, as in the Dominican Republic, trace directly to the original Spanish cattle brought to Hispaniola from Spain and the Canary Islands on the flotillas which immediately followed Columbus' discovery voyage:

This cow, seen in the Croix des Bouquets cattle market, was offered for sale as a dairy replacement. Haiti, 1970.

A cow on pasture in a dairy herd outside Port au Prince, Haiti, 1970.

A Criollo draft span on a large sugar plantation west of Port au Prince. The animal on the left is a bull. Such use of bulls under yoke is common on the island. Haiti, 1970.

not introduced into Haiti until 1949, and the few representatives of the northern European dairy breeds came in even later.

The Haitian Criollo is small, a cow in fair condition weighing from 600 to 800 pounds, the variation being largely due to the quality of the pasture she grazes. Bulls are 100 to 200 pounds heavier. Again, the most frequent color is Jersey-tan, usually of the light shade. Black, sometimes with white markings on the underside, and mottled black and white are less

206

common. The horns are not as large as on most Criollos possibly due to a mineral deficiency. Other features are true Criollo—thin tail, small brush, rounded barrel, and narrow, muscular face.

The cow in Haiti never received the attention that her sisters were accorded in some areas of comparable environment. Lack of breeding control, together with nutritional deficiency, probably account for her small size. Except in the dry areas there are fair grass lands of the type common to the tropics. But the high density of the human population has crowded even the subsistence farmer so that the cattle are usually closely confined, and roadside grazing is common.

The subsistence farmer holds most of the national herd. With the exception of a couple of 100-cow dairies using imported bulls and a few large plantations that keep Zebu bulls for the production of draft oxen, the cattle of the country in the early 1970's were still pure or nearly pure Criollo. The owner of a few cows sometimes obtains the service of an American Brahman bull or of a crossbred son. If he is near one of the large dairy farms, he may get his cow bred to a Holstein-Friesian or Brown Swiss bull. Eventually the crossbred criollo will replace the pure Criollo in the country but not before several cattle generations have elapsed. Complete grading up to the Zebu or to a dairy breed appears to be much farther down the road. Thus the breeding out of the Criollo does continue, but progress is slow either because imported bulls are not available or the subsistence farmer lacks interest in improving his cattle.

The Cuban Criollos that trace back to the original Spanish introduction have survived to the present day—maintained since the 1959 revolution by the state farms. The 30-year war for independence from Spain, which began in 1868, was probably the reason the Criollo in Cuba was not interbred with the Zebu to the extent that occurred on other islands of the Indies. The landed proprietor became the backbone of the

Cuba

207

The Cuban Criollos are the largest descendants of the original Spanish cattle. Their progenitors reached the island in 1511. These representatives were at the Granja Papi Lastre, Oriente Province:

This five-year-old bull, weighing 2,050 pounds, shows true Criollo characteristics. Oriente Province, Cuba, 1972.

A twelve-year-old cow weighing 1,200. She had recently completed her tenth lactation. Her record was given as 13,200 pounds of milk in 264 days. Oriente Province, Cuba, 1972.

resistance and the disruption of normal plantation life gave him little time for his livestock.

The excessive slaughter and lack of attention to replacements during the struggle left Cuba woefully short of cattle. When hostilities were concluded in 1899, efforts to replenish the national herd resulted in cattle being shipped in from any available source. Mexico was the largest supplier, but Colombia, Puerto Rico, and Florida also contributed. While much of the imported stock went to slaughter, a part entered the breeding herds. To what extent the Criollo was diluted with other types at this time will never be known. The areas from which the importations came had mostly pure Criollo stock at the time. All factors considered, it can be concluded that such additions as entered the Cuban national herd at the opening of the twentieth century did not markedly influence the genetic composition of the island's cattle.

When conditions stabilized in the early 1900's, Zebu bulls were brought in from Colombia and Venezuela and, later, from Mexico. These were crossed on the Criollo in some areas to obtain a better draft animal for the cane fields. Around 1910 Charolais were imported from France, and upgrading to the French breed began. Other northern European beef breeds followed and crossbreeding was practiced on many plantations. In some areas, particularly in Oriente Province, many Criollo herds remained uncontaminated. By the late 1940's criollo and pure Criollo cattle are said to have comprised 90 per cent of the total cattle population on the island.

The Criollo had certainly undergone some selection for an improved draft type long before the hostilities with Spain began. The landed proprietors had a heritage of cattle management going back to the open-range days of the hatos. Some attention was also probably given to the production of the haciendas' dairy herds. During the long years of the war, however, little attention was given to breeding practices. When mechanization began on the cane plantations after World War II, the selective breeding practices turned to improvement for beef and milk production.

The troubled days of the 30-year war with Spain, the heavy importations of breeding stock in the early 1900's to revive the decimated herds, the trend to upgrade to a Zebu draft animal—all are factors which have a bearing on the genetic composition of the Criollo as seen in Cuba today. Basically, however, this animal is the product of three centuries of natural selection followed by 150 years of intermittent planned breeding.

The color of the Cuban Criollo is remarkably uniform wherever it is seen on the island. The Jersey-tan color, sometimes with a reddish tinge, is solid over the entire body with a tendency to a lighter underline and extremities. There is some evidence of Zebu influence—horns are short and thick, unlike the widespread horns of the old Spanish cattle. (The practice of calfhood dehorning of cows is now general.) The sheath, umbilical fold and dewlap frequently indicate Zebu breeding, but there is no evidence of the hump on either the cow or bull.

The Cuban Criollo is the largest nonhumped breed in tropical America; in well-managed herds, mature cows weigh 1,200 pounds and bulls 1,800 pounds or more. Herds on test have averaged 6,160 pounds of milk for a 244-day lactation, with some individuals recording up to 13,200 pounds of 4.5 per cent butterfat milk. Conformation is of a dual-purpose, milk-beef type for which the state farms are now selecting.

In 1972 the total number of the breed on the island was estimated to be 80,000 head, in a total cattle population of 7,600,000. Present indications are that this descendant of the old Spanish cattle will continue into the foreseeable future.

Part IV. FULL CIRCLE

The trails followed by the Spanish cattle and their descendants have now been traced from Andalusia and the Canary Islands to the Indies, on to the continents, and down to where the scattered remnants of their issue, the Criollo, now remain. The chronicle extends over the past five centuries. It is now time to return to the lands of their origin in Andalusia and the Canaries and view the lineage of the old Spanish cattle which did not cross the ocean seas.

The Spanish breeds in Andalusia and the de la Tierra cattle in the Canary Islands, as seen today, are presented in the following pages. They will then be compared with the Criollo breeds of the Western Hemisphere.

Cattle in Andalusia

The Retinto is the predominant breed in Andalusia today as it has been for countless generations. There are scattered herds of the Black Andalusian and the black-eared Berrenda breeds, but these are rapidly fading out. The pure white Cacereño, native to southern Estremadura, is nearly extinct. It is a logical as-

213

The large herd of Retinto on the Las Lomas Finca, near Casas Viejas southeast of Cádiz, are known locally as "Tamerone" Retintos. Selection has produced a type larger than is typical of the breed:

This mature bull weighed at least 1,800 pounds in a very dry year. Casas Viejas, Spain, 1975.

A mature cow that probably weighs 1,200 pounds. Casas Viejas, Spain, 1975.

sumption that for the past five centuries all of these breeds, and their progenitors before them, have enjoyed the hospitable environment of southwestern Spain—in the same general areas where they are now found.

These different types have been maintained under levels of management that have advanced over the past 500 years in about the same manner as has stock raising since 1880 on the western plains of the United States. The open range ranching, large herds with little control over breeding, gave way to fenced pastures, smaller holdings, varying degrees of breeding control, and eventually some selection for a better beef animal. The introduction of foreign breeds reached significant proportions only a decade ago. And while crossbreeding has taken a severe toll of the old breeds, true representatives still graze the Andalusian pastures.

Retinto The drainage of the Guadalquivir River from Córdoba to the coast has been the homeland of the red cattle, the Retinto, as far back as there is any record. There is considerable variation in size and shade of color in the populations in different areas, but other characteristics of the Retinto hold true to type wherever it is seen. Varying levels of management, the quality of pasture, perhaps in a certain locality the owner's preference for a particular color shade—all account for these differences.

Color is typically solid over the entire body and varies in different herds from a Jersey-tan to a bright cherry red. This variation in color is no greater than that of the light and dark reds which Hereford breeders have fostered in their animals at various times and have accomplished within a few generations. Any white on the Retinto is frowned on, but the extremities are usually of a lighter shade.

The horns on the cow are rather thin, very wide, upspread, and frequently have lyre-shaped ends. On the bull the horns usually are much thicker and shorter, but on some individuals widespread horns upturned at the ends are seen. The head is long and narrow. Conformation, while indicative of a beef

216

type, is inclined to be rather rough. The legs are large boned, and the body stands well off the ground. Top and bottom lines are quite parallel; there is little tendency for the bottom line to cut up at the flank. The barrel is well rounded but not particularly deep. Mature cows in some areas near the coastal plain average 1,000 pounds, and bulls 1,500 pounds. In other areas the average cow will weigh up to 1,200 pounds and bulls 1,800 to 2,000 pounds.

Many Retinto herds have recently been crossed with northern European breeds, particularly the Friesian and Charolais. There has also been some crossing with the Santa Gertrudis. While this practice is increasing, many Spanish cattlemen continue to maintain pure Retinto herds. The population, though gradually decreasing, appears to be in no immediate danger of extinction. The Retinto Breed Society was founded in 1965. This is something of a rarity in this country where there are only five bovine registry associations and herdbooks.

After man learned that he could predetermine to a degree the characteristics of his animals by selection of their parents, such nonutilitarian features as horns or color patterns were often among the principal objectives of his breeding programs.

Berrenda

During medieval times in Spain a cow with a white body and black ears must have appealed to some cattle owners in the hills around Córdoba. Such conjecture has not been recorded, but these white cattle with their black (or red) ears have existed for countless generations in northern Andalusia. The Berrenda has a predominantly white body commonly marked with minor black spots around the shoulders and on the neck. Occasionally solid black patches are seen. The ears are invariably black or red. The recessive red gene, as in the black and red Angus, accounts for red markings entirely replacing the black in some herds.

The Berrenda is quite similar to the Retinto in general conformation. The average weight of mature cows is around 1,100 pounds. Bulls weigh 1,600 to 1,700 pounds. Horns on the

A Berrenda cow on an old family finca outside Córdoba. An exceptionally dry season accounts for her poor condition. The individual feed bunk of hewn stone at the left is a relic from the Roman occupation. Córdoba, Spain, 1975.

cows have the pronounced wide spread typical of Spanish cattle; the horns on the bulls are shorter and heavier.

The homeland of the Berrenda is not sharply defined. The breed has been known for many generations in local areas of northern Andalusia and adjoining provinces to the north. A few individual herds are still maintained pure in Córdoba Province—the continuance of old family tradition or the love

218

of an old-time cattleman for his animals. In a few more generations the breed could easily disappear.

The Cacereño, the all-white breed described below, is said by Spanish cattlemen to have no ties to the Berrenda, a supposition that may prove to be in error. The only distinctive differences between these two types is that the Cacereño has neither black spots on the body nor black ears. Several generations of selection for a solid white hair color could probably eliminate the black spots on the Berrenda's shoulders and the black ears. Selective matings of black-ear carriers could also introduce the black-ear feature into a herd of all-white cattle. The possibility must be recognized that the Berrenda could have been so derived from the Cacereño in the distant past. Future blood-type investigations may some day furnish the answer.

That no record can be found of white cattle with black ears being shipped to the New World is not particularly significant—lack of any description of the bovines that stocked the Indies has frequently been mentioned. But the marked similarity of the Berrenda in Spain and the Blanco Orejinegro in Colombia is practically irrefutable evidence that Berrenda cattle, in some unrecorded manner, did reach the New World.

The provinces of Cáceres and Badajoz are said to have harbored the Cacereño cattle for centuries. These provinces are in Estremadura, which borders Andalusia in the northwest. No record has come to light of cattle from Estremadura being shipped to the Indies. The area, however, is the nearest of any known habitat of white cattle to the ports from which shipments were made. The Cacereño is mentioned here because white cattle, other than the Berrenda, are occasionally seen in descendants of the Spanish cattle in the New World.

Cacereño

Local folklore has the white ancestors of the Cacereño being brought to the Guadalquivir Valley by the Romans when they began their colonization efforts in the second century before Christ. The Romans in their day on the Iberian Peninsula

The Cacereño breed is nearly extinct. These representatives were in the 30-head herd collected in the early 1970's by the Ministry of Agriculture in an effort to perpetuate this old type of cattle:

A mature bull at the artificial insemination station. Badajoz, Spain, 1975.

220

A cow with calf fed in dry lot because of the drought. Badajoz, Spain, 1975.

The Black Andalusian, now diminishing in number, can still be seen on the rolling plains of northeastern Andalusia:

A young bull, weighing 1,200 pounds. Córdoba, Spain, 1975.

A mature cow on sparse pasture. She was in poor condition as a result of the 1975 drought. Córdoba, Spain, 1975.

certainly husbanded cattle and gave serious attention to their care. The individual feed bunks in which they fed their cattle are still in use on a few old family fincas. These bunks were built to last—a cube of solid stone, hewn by a Roman mason, with a bowl-like basin chiseled in the top to hold grain. Drainage was provided by a hole drilled through the stone to the bottom of the bowl. Some future delving in the archives, however, will have to authenticate the movement of white cattle from Italy to Spain. All-white cattle and white cattle carrying black ears must have grazed the pastures of the Iberian Peninsula in the fifteenth century, but where they came from has not been disclosed.

An effort to perpetuate the Cacereño is being made by the Ministry of Agriculture. Thirty cows and two bulls were selected as the best representatives obtainable from the few remaining small herds. These are being maintained under carefully controlled conditions by the artificial insemination center at Badajoz. The total number of the Cacereño now in existence is said to be fewer than 100 head.

Black Andalusian

The slopes of the mountains of south central Spain and from there down to the Andalusian plains were the homeland of the Black Andalusian cattle. These are solid black animals. The horns on the cow are widespread and upturned, not as large as on the Retinto but of the same typical shape as most Spanish cattle. Mature cows weigh around 1,000 pounds, bulls 1,500 pounds. The rather thin face, prominent tail stock, and the thin dewlap are characteristic. The Black Andalusian has a draft-type conformation, but is now considered a beef breed.

Unless its habitat has shifted at some time in the past the Black Andalusian would have been available for shipment down the Guadalquivir River in Columbus' time. Only scattered herds of these black cattle are seen in the Guadalquivir Valley today. Crossbreeding to the northern European breeds has taken its toll and still continues. Black animals are still seen in herds of Criollo cattle in widely separated parts of the

Western Hemisphere where there appears to have been no introduction of non-Spanish cattle. Black cattle have been noted in early colonial days in New Spain. This is the most convincing evidence that some progenitors of the Black Andalusian were included in the early shipments to Hispaniola.

The fighting cattle have been raised as a distinct type in Spain longer than any recognized breed in Europe. Their written pedigrees treasured by some breeders go back to the eighteenth century. Breeding fighting cattle continues in carefully segregated herds throughout Andalusia, and the type is in no danger of extinction.

Fighting
Cattle

The incredibility of the De Lidia contributing to the several cattle populations of Latin America has been discussed. They were, however, introduced to the Americas—to New Spain in the sixteenth century, to Colombia at a much later date, and continue to be raised in both countries. The fighting stock in the Americas has descended directly from the best herds in Spain by means of many importations, particularly in later years. Throughout the intervening centuries the same breeding pattern has been followed on both sides of the water, and the fighting cattle have had no contact with the common cattle of the country. These idealized bovines have followed the course of the Spanish empire along a line parallel with that of the Criollos, but their paths have never crossed.

De La Tierra Cattle in the Canary Islands

The Canary Islands had no indigenous cattle. Breeding herds were established by the Spaniards as they subdued the Gaunches and began colonization in the last quarter of the fifteenth century. The conquest of the islands was the preliminary move in the western expansion of Spain that continued for 300 years.

Small dairy herds of de la Tierra—cattle of the country—can still be found in the mountainous center of the Grand Canary Island. These cattle bear a marked resemblance to the Retinto of Andalusia. Both descended from the same progenitors in fifteenth-century Spain. Until recent times the de la Tierra were the only cattle in the Canary Islands:

This mature bull, one of the last on the island, weighed 1,800 pounds. He had been purchased from a farmer for use in a tourist attraction—"a wild west show"—to take the part of a Longhorn. Grand Canary, 1975.

This cow was a member of a three-cow dairy in the mountains. Her production was placed at 10 liters per day after her calf had sucked. Grand Canary, 1975.

Here the pattern was set for giving land grants, which carried the services of the inhabitants, to the conquerors after resistance of the native population had been broken. Cattle—to pull the plow and furnish meat and hides—were introduced so that the settlers could become self-supporting. The most important use of cattle initially was for draft, as wild hogs, sheep, and goats were available on all the islands for meat. Only two decades elapsed between the time the first breeding herds were established in the Canaries and when their descendants were being moved out to the Indies to found breeding herds there.

Materials and supplies for colonization of the New World were generally loaded at the ports of Sevilla and Cádiz where all articles of commerce, as well as livestock, were readily available. All ships routinely stopped at ports in the Canaries which was an established way station on the route to the Indies. Cattle were taken on here, probably to the limit of the islands' ability to supply them. The advantages over loading livestock at mainland ports were a shorter trip, fewer losses, not as much feed to be carried. The repeated orders of the *Casa de Contratación* to shippers to take on cattle at the Canaries instead of mainland ports, and specific requests of the colonists for island cattle, appear at times to have been ignored. This is good evidence that the Canary cattle were not always available. The demand was invariably for pregnant heifers and bull calves in order to minimize the space needed aboard ship for livestock. Such animals would have been more readily available around Sevilla and Cádiz, which were surrounded by a much larger bovine population than the island ports. A very substantial part of the cattle that stocked the Indies did originate from the Canary Islands, but how the number compared with that leaving from Andalusian ports is not known.

There is no evidence that any cattle, other than those from Andalusia, were brought to the Canary Islands until recently. In the course of time the inhabitants came to consider their cattle as native to the land and called them *de la Tierra*. They were bred for draft and milk animals, and down through the years there was undoubtedly rudimentary selection for these traits.

228

All of the islands have high mountains in the center, and there are wide variations in rainfall as well as in the character of the terrain. These environmental influences over the five centuries that have elapsed since cattle first reached the Canaries have produced some variation in the type of cattle that evolved on individual islands. Such differences do not mask the broad similarity between the de la Tierra and the Retinto; that both have descended from the same fifteenth century progenitors in Andalusia is obvious. There is no evidence that either the black or the white-bodied, black-eared cattle of Spain ever reached the Canaries.

The de la Tierra has now nearly disappeared on all the islands. Widespread use of the tractor for farm work and the truck for general hauling have almost eliminated the draft ox. The demand for milk created by the tourist trade has seen the de la Tierra bred up to the more productive European Friesian. By the mid-1970's, true descendants of the old Spanish cattle could be found only in remote areas on any of the islands.

Grand Canary

The Grand Canary, half the size of Rhode Island and the most heavily populated island, has a small number of excellent de la Tierra cows in the mountainous central area. Because of the high altitude this region enjoys a better than average rainfall. Hillsides and rocky areas not suited to cultivation provide good pastures for part of the year. Small dairies, with 2 to 6 cows, make good use of this land. A few of these farms still hold on to their de la Tierra cows although their neighbors have usually gone to Friesians. A cow of the old breed that has just freshened invariably has a Friesian-sired calf; the bulls have practically disappeared.

The color of the de la Tierra on the Grand Canary is a Jersey-tan with a reddish tinge. Cows weigh up to 1,100 pounds. Conformation bears a marked resemblance to the Retinto as also do the solid color and long face. The horns, however, are not as large as those of the Andalusian cattle.

Selection for a draft type is evident in the de la Tierra cattle that remain on the Grand Canary and Tenerife islands:

The de la Tierra draft cow continues to be worked on a few farms on the island of Tenerife. The type is now practically extinct. Tenerife, Canary Islands, 1975.

A de la Tierra ox span on the Grand Canary Island. The ox on the left weighed 1,300 pounds, the one on the right 1,250 pounds. Oxen are used to cultivate odd corners of the terraced fields. Grand Canary, 1975.

The Grand Canary, along with Tenerife, is more progressive than the other islands, but the draft ox still finds his place in the field. Hillsides are heavily terraced for the utmost utilization of the land, and there are corners too sharp for the maneuver of a tractor. It is here the de la Tierra ox team still draws the plow.

Tenerife   On Tenerife the de la Tierra milk cow seems to have disappeared. A few of her sisters still remain, but it is obvious that their lifetime has been spent under the yoke. These are lean and muscular animals, a solid brown color over the entire body except for lighter extremities, their horns curved to the shape of the yoke. Old residents say their cattle have always been brown in color. The average cow in fair working condition weighs 1,200 pounds. Bulls can no longer be found.

In the early days of shipping to the Indies, Tenerife is not mentioned as a port of call. It was not secured until 1496, the last of the islands to be conquered. Probably no cattle were available for export until after Hispaniola had been adequately stocked. The brown color of the few surviving Tenerife de la Tierra cattle poses an unanswered question. In no other place where descendants of the old Spanish cattle can be seen has the pronounced brown of the Tenerife animals been noted. A possible local preference for this color and some selection for it by growers on the island are the only apparent explanations. In the few survivors that can be seen today there is no indication that any cattle type, other than the progenitors of the Retinto, could have had a place in their ancestry.

Gomera   Gomera must have been one of the early sources of cattle for the Indies. It was one of the ports routinely made by Columbus, where he was rumored to have had a romantic interest. Gomera was the only island where the Portuguese had subdued the natives when Spanish possession was acknowledged in 1479. Thus secured, it would have been the first to be stocked with breeding animals from Andalusia. The island is only 145 square miles in area, and the potential for raising cattle was obviously limited. After filling the early demand for the Indies, the supply of animals for export must have soon been exhausted.

The de la Tierra cattle now on Gomera are smaller animals than those seen on the other islands. The best cows weigh around 900 pounds. Color is the typical Jersey-tan to a light red. The horns are much smaller than those usually seen on Spanish cattle. There is no apparent explanation for the smaller size and smaller horns on the Gomera cow. A nutritional deficiency might be suspect, but Gomera is of the same volcanic origin as the other islands and should have the same type of soils. Rolling hills occupy much of the terrain and must have provided excellent pasture if not overstocked. The small size of the Gomera cattle remains another unanswered question in the history of the de la Tierra.

La Palma could not have been an early source of cattle for the Indies. The island had only been conquered the year before Columbus sailed on his discovery voyage. The broken country north of the central mountain ridge is good cattle country, and the ancestors of the Retinto must have multiplied rapidly there as soon as they were introduced. La Palma is frequently mentioned as a port of call, and in the later years of cattle movements to the Indies must have been an important supply point. **La Palma**

Today, on the rolling hills at the north end of La Palma, there are more de la Tierra cattle than can be found in all the other islands. A few farms have herds of ten or twelve cows. Young stock is grown out on grass and then sent to the local abattoir for beef. Individual dairies have four to six head. There is no trouble in locating a bull, kept by a grower for his own cows as well as to provide service for his neighbors.

The color of the de la Tierra on La Palma varies from a light tan having a yellow cast to a reddish Jersey-tan, and is always solid over the body. The horns on some individuals will rival those on an Andalusian Retinto but normally are considerably smaller. The average cow in good condition weighs 1,100 pounds. Only young bulls are now kept, and at

233

Considerable variation in such characteristics as size, shade of color, and shape of horns is seen in the de la Tierra on the individual islands. Such differences, which have been determined by varied environments and methods of management, do not mask the common heritage of all the Canary Island cattle:

The de la Tierra milk cow of a village dweller on Gomera Island. She weighed around 850 pounds. The de la Tierra there were much smaller than those on the other islands. Gomera, Canary Islands, 1975.

234

A typical de la Tierra cow on the island of La Palma which held the largest population of this old-type cattle of all the Canaries. La Palma, Canary Islands, 1975.

two years of age will weigh about the same as a mature cow. When around two years of age, bulls are sent to market and replaced by a yearling. This practice increases the return from the abattoir and cuts down on the consumption of feed.

La Palma is the only island having a number of de la Tierra cattle that is large enough to furnish a nucleus for preserving the breed. No effort along this line appears to be contemplated. The Spanish Ministry of Agriculture has authorized programs to perpetuate some of the old Peninsular breeds. The founding of the herd of Cacereño at Badajoz with this objective has been mentioned. The gene pool of the de la Tierra on La Palma could be saved from extinction by similar action if this were taken quickly.

Spanish and Criollo Breeds Compared

The Spanish cattle in Andalusia and the widely varied types of Criollo in the Western Hemisphere present an intriguing profile reflecting the effects of environment and man's management. The saga began in the late fifteenth century when those few hundred head of Spanish cattle were transplanted to the Indies. The founding herds in Andalusia and the Canary Islands continued grazing on their native pastures down to the present day. Their herd mates that went to the Indies propagated beyond all expectations and were dispersed to the varied climates of the Western Hemisphere.

The Castilian cattleman of the fifteenth century continued his open-range ranching in Andalusia until the increase in human population inevitably led to smaller holdings and fenced pastures. Under these more intensified conditions, cattle were bred with an eye to selection for quality and type. Much the same pattern of husbandry proceeded in the Canaries on a smaller scale.

Only recently were the foreign dairy breeds brought to the

Canary Islands and the de la Tierra cattle bred out; only a few scattered remnants remain today. The decline of the old Spanish breeds in Andalusia began even later but progressed rapidly. Two of the smaller breeds, the Berrenda and the Cacereño, have nearly disappeared, but many good herds of the major Andalusian breed, the Retinto, are still maintained. The Black Andalusian has also been reduced in numbers, although scattered herds can still be seen in the province of Córdoba.

The open-range methods of management in Andalusia followed the Spanish cattle as they spread throughout the Americas and were generally continued for three centuries. The first cattle in the Indies found better pastures than they had grazed at home. The island grasses were probably not as nutritious as those in Andalusia and the Canaries, but the supply was unlimited and the pastures were green throughout the year. The climate was warmer but not oppressively so. The rapid increase once the foundation stock became established is ample evidence of adequate nutrition and an environment to which the Spanish cattle adapted readily.

When herds were developed on the mainland, nutritional deficiencies, heat stress, high altitudes—all the unfavorable environmental influences which plague cattle—were encountered in some areas; in others, a bovine Utopia was found. Man's management of his cattle also varied widely, from a complete lack of breeding control on the llanos to a selection for a draft or dairy type at a later day in more advanced areas. Serious mineral deficiencies in Florida resulted generations later in the Scrub. The poor quality of the llanos grasses and the long dry season forced a rapid adjustment of the Spanish cow to such unfavorable conditions. When she reached the Cauca Valley in Colombia, in the warm temperate climate and on good quality grasses, she was in cow heaven. The pampas of the Argentine and the Uruguayan grasslands ultimately proved to be the best cattle country of all Latin America. The high plateau of New Spain was also hospitable to the Spanish cattle. In the mountainous regions of Central America a smaller animal evolved than on the coastal plains.

237

Thus the Spanish cow spread out in the various environments that the Western Hemisphere had to offer. She adapted herself wherever man led her, and in three centuries made herself at home in all parts of the Americas, displaying wide variations in size, conformation, and productivity as the environment dictated. With it all, she responded well to whatever selection practices man employed.

On the following pages, illustrations of Spanish cattle in Andalusia today are placed in juxtaposition to five of the modern Criollo breeds in Colombia. These close the circle between the Spanish cow and the Criollo after five centuries of divergence. The red breeds—the Retinto in Spain, and the Romo Sinuano, Costeño con Cuernos, San Martinero, and Chino-Santandereano in Colombia—have been subjected to environments not radically different and with management practices of comparable patterns. The same can be said of the white breeds—the Berrenda of Spain and the Blanco Orejinegro of Colombia. Here, as has been noted, there is the missing link as to when and how the progenitors of the Berrenda got to Colombia. There is an unexplained difference in the shape of the forehead of the Berrenda and the Blanco Orejinegro which is in conflict with the similarity of other features.

After nearly five centuries of development on opposite sides of the ocean sea, the tie between the Spanish and Colombian breeds is inescapable.

Spanish and Criollo Breeds Compared

Retinto bull, near Córdoba, Spain, 1975.

Costeño con Cuernos bull, Granja Turipana. Montería, Colombia, 1975.

Romo Sinuano bull, Granja Turipana. Montería, Colombia, 1975.

Retinto cow. Casas Viejas, Spain, 1975.

Costeño con Cuernos cow, Granja Turipana. Montería, Colombia, 1975.

Romo Sinuano cow, Granja Turipana. Montería, Colombia, 1975.

Retinto bull. Casa Viejas, Spain, 1975.

Chino-Santandereano bull. Fondo Ganadero de Santander, Bucaramanga, Colombia, 1975.

Retinto cow (horns blunted). Casa Viejas, Spain, 1975.

248

San Martinero cow. Carimagua, Colombia, 1975.

Berrenda cow, near Córdoba, Spain, 1975.

250

Young Blanco Orejinegro cow, San José del Nus, Colombia, 1975.

The Criollo Under Natural Selection

The preceding illustrations of contemporary Spanish cattle and Criollo breeds in Latin America show their high degree of similarity. But there are tangents to the circle. These are the Criollo cattle segregated out of the common gene pool by differences in environment and management. These variants, such as the Llanero, the Casanareno, and the Criollo of Central America, true line descendants of the old Spanish cattle, are the product of natural selection in environments that differed markedly from that of their progenitors in Andalusia.

Color, horn shape, and body size are the attributes of the bovine which first catch the eye. All of these characteristics can be modified by artificial selection—the choice of the breeding animals man makes to continue a line of descent.

If the lightest-colored animals in a red herd are chosen for breeding, in a few generations a herd of light red cows results. Nature will accomplish a similar result, provided there is a survival factor involved, but over a much longer time interval. If a herd of red cows is moved to an area of extreme sunlight, those of lighter color may do better, suffering less from the heat because more of the sun's rays are reflected from their bodies; they live longer, produce a larger proportion of the replacements, and after many generations the entire herd takes on a lighter shade of color. When cattle have been moved from one climate to another, or from an area of adequate forage to one deficient in nutrients or essential minerals, wide differences in the horns as well as in body size have resulted from nature's forcing adaptation to the new environment. Such modifications of an animal's characteristics are natural selection.

In the past, both man and nature have induced radical changes in the Criollo cow in a diversity of environments. These changes appear to have attracted very little attention from the animal scientists and geneticists, and it is not the intention to resolve them here. Similarities and dissimilarities in color, horn size and shape, and body size of Spanish cattle and their relatives of the Western Hemisphere are noted on the following pages. The objective is to present the different types and breeds as they exist—not to explain their evolution.

There are differences in size and shape of horns between the Spanish cattle in Andalusia and the Criollo types and breeds in the Western Hemisphere. The Costeño con Cuernos of Colombia and the Criollo cow in Haiti display sets of horns comparable to those of the Retinto in Spain. In the Latin American countries, where the Criollo has been subjected to artificial selection, dehorning of breeding animals is now common practice, and comparisons of horn size and shape are difficult to make. A cow or bull with horns is rare, and when the odd one is seen, the question arises of how representative it is. **Horns**

In general, the horns of the Criollo are neither as large nor as widespread as those of present-day cattle in Spain. Horns on the cows in small herds that have had no artificial selection are usually larger than might be expected when the size of the animal is considered. In areas of habitually deficient feed, the horns are usually small.

The Longhorn is a special case. Being a hybrid, it does not display all the pure Criollo characteristics. Also, aged steers constituted a majority in the cattle drives out of Texas. Loss of the male hormone by castration can markedly increase horn size in an aged animal. And it was these old steers that earned the Longhorn its name and reputation. Over the past fifty years selection for prominent horns has been made in the refuge and fancier herds that are maintained in Texas and Oklahoma. To what extent this has influenced horn size is problematical, but the Longhorn today commonly carries horns larger than seen on the Spanish Retinto.

The size of the Criollo ranges from the 500-pound Casanareno cow in the Colombian llanos to the 1,200-pound Cuban Criollo. Differences in environment, nutrition, and management easily account for such variations. **Body Size**

253

Where the Criollo has been under good management and maintained on adequate feed, as in the herds in Colombia, Cuba, Venezuela, and Costa Rica, size is comparable to its overseas relatives. The Florida Scrub, the Chinampo, and the small Criollo in the llanos or in the hinterlands of Middle America have been on deficient feed for centuries and have been subject to no breeding control. Their small size is the obvious result of environmental influences.

Color A common misconception as to the color of Criollo cattle must be corrected. Modern writers sometimes refer to the old Spanish cattle, or their progeny, as "extremely varied in color," or as being "all colors of the rainbow." Barring the blacks and whites, which were probably only a small minority of the foundation herd on Hispaniola, the Spanish cattle appear to have been predominantly of a solid color. This varied in intensity from Jersey-tan to cherry red and occasionally brown. This conclusion is supported by authoritative writers who occasionally mentioned the color of early Spanish cattle. A number of such references follow:

> *1511*: Description of the cattle taken from Hispaniola in Columbus' time: "The original cattle were of a type familiar to Spain in the 16th century; rather small, well-shaped and handsome animals, of a light brown or dark jersey color, similar to that of the wild deer in shade, and usually carrying a dark streak along the spine, with a rather heavy cross of black at the shoulders." (Johnson, *History of Cuba*) It would be interesting if Johnson, who was writing in the early 1900's, had disclosed the source of this description of fifteenth-century cattle, but he failed to do so.

> *1540–41*: Coronado in his writings referred to the bison which he encountered north of the Rio Grande as "oxen," being unfamiliar with this North American animal. He reported: "These oxen are of the bigness and *color* of our bulls." (Cox, *Cattle Industry of Texas*) The significance is in the wording "color of our bulls" employed to indicate the color of Spanish cattle in Coronado's time.

1799: Quoting a traveler through Attakapas County in Louisiana: "We met a herd of about 1,000 head . . . they looked more like deer than cows and oxen . . . their usual red-brown color heightened the illusion." (Post, *Cattle Industry of Louisiana*)

1805: Describing cattle on the plains of Venezuela: ". . . very uniform in color." (Humboldt, *Equinoctial Regions*) No designation of the color is given.

Early nineteenth century: Description of the cattle in southwestern Louisiana: "They were *brick colored stock* with immense wide spreading horns and were said to be descendants of the original Spanish stock." (Gray, *History of Agriculture*)

1820: Referring to an early search of cattle records in Colombia and Venezuela: "In all the country where criollo are found, red, or tawny red, or black and white spotted predominate." (Pinzón, *Ganados Nativos Colombianos*)

1851: Captain Richard Ware reported the wild cattle he saw in Texas as ". . . all one color—brown." (Dobie, *Longhorns*)

1901: The Criollo cattle of Puerto Rico were described as ". . . fawn and dark colored jerseys" for many animals and ". . . some have a pale, and others a rich colored Guernsey color. Black, white spotted, dark red and roan cattle are not uncommon. . . ." The mention of "roan" can be questioned. None of the other writers refers to a roan color. (*Annual Report of Bureau of Animal Industry, 1901*)

Such evidence from old writings that has come to light clearly indicates that some variation of a red color was most frequently observed. There are also credible references to "black" cattle, but rarely in a context where it is clear that an animal actually black in color was being described.

The breeds or types of Criollo which have undergone a period of artificial selection in widely separated areas of the Western Hemisphere all show a significant similarity in color as well as in many other characteristics. These types, the areas, and the pages on which they are described are shown in the accompanying table.

THE CRIOLLO BREEDS TODAY

Types	Area	Page
Dairy		
Milking Criollo	Costa Rica	179
Limonero	Venezuela	128
Chino-Santandereano	Colombia	140
Vallecaucano	Colombia	137
Costeño con Cuernos	Colombia	146
Criollo (milk type)	Mexico	167
Cuban Criollo (dual purpose)	Cuba	207
Draft and Beef		
Romo Sinuano	Colombia	150
San Martinero	Colombia	141

All of these breeds or types display the solid Jersey-tan to cherry red color, and all underwent a period of selection by man for some definite trait, such as milk production, draft capability, or beef conformation. So far as the history of these types can be ascertained each one was developed from the local descendants of the old Spanish cattle in the area where it now exists.

The color patterns of the Criollos which were not subjected to artificial selection are more varied than those which have been under breeding control by man. However, the Jersey-tan color, with few exceptions, predominates in any group of Criollos. Less prominent patterns are the solid black and the black-and-white. A moderate-sized group of Criollo cattle usually has a solid black animal or two. The black-and-white pattern may be quite varied—nearly solid white with mottled black and white around the shoulders, white with black spots over all or part of the body, or a mottled black-and-white pattern over the entire body. The white body and black ears of the Blanco Orejinegro is far less common but occasionally is seen in groups of Criollo far removed from their habitat in the Cauca Valley of Colombia. Brindles are rarely seen, and large black-and-white patterns, similar to that of the Holstein-Friesian, seem to appear only when there has been interbreeding with other cattle types.

The illustrations on the following pages were chosen to show the similarity in the color patterns of the Criollo as seen in widely separated regions of Latin America. Differences in body size, shape of horn, and physical condition, that have resulted from natural selection in widely varied environments, do not mask the broad conformance to the typical Criollo type. Many of the photographs are of individuals in countries where the Criollo has become nearly extinct. Each animal photographed, however, was seen under circumstances which would fairly substantiate its being a pure, or nearly pure, descendant of the old Spanish cattle.

Jersey-tan Criollo:

Cow from a large herd in the Beni, Bolivia, 1969.

Jersey-tan Criollo:

Tethered cow of a small farmer. Antigua, 1970.

Jersey-tan Criollo:

A Casanareno cow in the llanos, Colombia, 1975.

Jersey-tan Criollo:

Cow of a mountain dairy herd, Honduras, 1969.

Jersey-tan Criollo:

Cow of a small grower in Trinidad, 1970.

Jersey-tan Criollo:

Llanero cow near the Orinoco River, Venezuela, 1966.

Jersey-tan Criollo:

Cow in a small herd in Surinam, 1970.

Jersey-tan Criollo:

Farmer's cow that had been brought to the government experiment station for artificial insemination. French Guiana, 1970.

Jersey-tan Criollo:

Cow in a village dairy herd near the north coast in the Dominican Republic, 1970.

Jersey-tan Criollo:

Cow in the Croix des Bouquets cattle market, Haiti, 1970.

Jersey-tan Criollo:

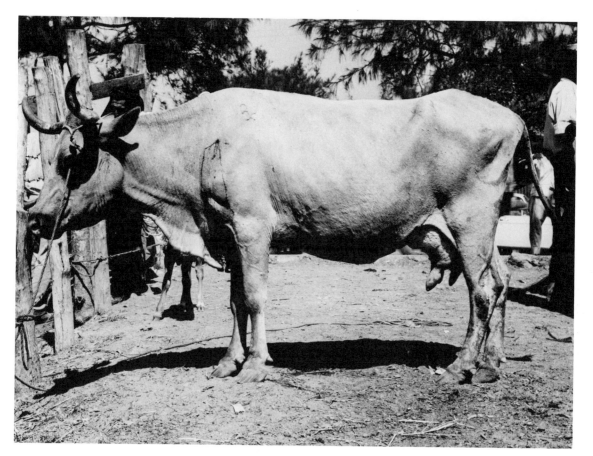

Cow at the San Salvador cattle market in El Salvador, 1969.

Jersey-tan Criollo:

Cow in the eastern mountains of Guatemala, 1969.

Black Criollo:

Florida Scrub cow, in fair condition, with horns blunted, a type of Criollo that is now practically extinct. North of Lake Okeechobee, 1971.

Black Criollo:

Chinampo cow in Baja California, south of La Paz, 1971. The animal was in average condition for these Criollos which know only a desert pasture.

Black Criollo:

Criollo cow of a small farmer in southeastern Guatemala, 1969. The black color is not common among the few Criollo survivors in Guatemala.

Black Criollo:

A roadside-grazing Criollo cow in Haiti, 1970. The bar hanging from the neck was to prevent the free-roaming animal from entering enclosed cultivated plots, such as that seen in the background.

Black Criollo:

Criollo bull in the mountains northwest of Tegucigalpa, Honduras, 1969, an area in which there has been little crossing with foreign breeds.

Black Criollo:

Casanareno cow in the corral of a finca on the Casanare River, Colombian llanos, 1975. An odd black animal will usually be seen in a sizable herd of these Criollos.

Black and White Criollo:

A Criollo bull in the Beni of eastern Bolivia.

276

Black and White Criollo:

This Longhorn cow in the Wichita Mountains Wild Life Refuge near Cache, Oklahoma, displays the black-and-white color pattern that is occasionally seen in this herd. The black ears are reminiscent of the Blanco Orejinegro in Colombia. 1971.

Black and White Criollo:

    The Criollo steer in the holding yard of the abattoir at Condega, Nicaragua, was raised in the nearby mountains. 1969.

278

Black and White Criollo:

A Chinampo cow from the desert range brought into La Paz, Baja California, for shipment to the Mexican mainland. 1969.

Black and White Criollo:

A Criollo cow that was found in a small dairy herd in the mountains of northeastern Honduras, 1969.

The Criollo Heritage

The Spanish cattle multiplied rapidly and required little care from man in the subtropical climates to which they were introduced, and the increased population was expanded even more as straying animals "escaped" to a wild state. A century after the Spanish cattle had spread to both continents the pre-breed northern European cattle (to become known as the Native American cattle) arrived in the temperate zone of North America. These bovines increased at a slower rate, first because man's care was essential for their existence and, second, because the English colonists did not expand as rapidly onto the virgin lands to the west as did the Spaniards who had a whole hemisphere before them. These two cattle populations met when the human populations intermingled as the changing political boundaries permitted.

After 1800 the cattle population of the Western Hemisphere continued to grow essentially by natural increase. The composition of the herds was transformed by interbreeding, first, with bulls of the newly introduced northern European breeds and, second, with bulls of the Zebu breeds. The total number of imported cattle that effected this transformation could be counted in the thousands.

The cattle populations in the Western Hemisphere today and how they derived are indicated in Figure 3. The Criollo cow furnished the base for three-fourths of all these cattle. In the temperate zones the northern European breeds have taken over completely. North America, east of the Missouri River and north into Canada, saw the Native American cattle bred up to the northern European breeds in less than a century. On the great plains and west to the Pacific, the Longhorn, a near-Criollo that had moved up from Texas and Mexico, was converted to Hereford, Angus, and Shorthorn herds. Today a further conversion may be in the making, as the recently introduced European breeds enter the picture. On the high plateau of northern Mexico, the Criollo was upgraded to the Hereford and Angus.

LEGEND

GRADE AND PUREBRED NORTHERN EUROPEAN
CATTLE AND THEIR CROSSBREDS:

Upgraded from Northern European Pre-breeds

Upgraded from Criollo Cattle

ZEBU TYPES—GRADE, PUREBRED, AND
CROSSES DERIVED FROM CRIOLLO
CATTLE

ORGANIZED HERDS OF CRIOLLO EXTANT IN 1975

1 Cuba—Cuban Criollo
2 Haiti—Unimproved Criollo
        (unorganized but numerous)
4 Mexico—Improved Criollo
5 Costa Rica—Milking Criollo
6 Venezuela—Improved Criollo
7 Columbia—Blanco Orejinero
7a   —Vallecaucano
7b   —Chino Santandereno
7c   —San Martinero
7d   —Costena con Cuernos
7e   —Romo sinuano
7f   —Casanareno Criollo
8 Bolivia—Unimproved Criollo

BRAZIL
Mostly Zebu and
Criollo upgraded
to Zebu

Fig. 3. Distribution of the Founding Cattle Types in the Western Hemisphere, 1975

In the temperate climate of Argentina and Uruguay the Criollo vanished into the Hereford, Angus, and Shorthorn breeds within the past century. This paralleled the same pattern of conversion that had occurred in the western part of the United States. The tropics and sub-tropics, from Mexico to Paraguay and in the Caribbean Islands, became the land of the Zebu. Small pockets of pure Criollo have survived in several widely separated areas.

A number of sizable islands of northern European breeds, particularly the Holstein-Friesian, have been planted in the tropics, have acclimated to a degree, and are surviving satisfactorily. Such areas are usually found at relatively high elevations where temperatures are lower and heat stress diminishes. In the highlands of Central Mexico, the sierra of Ecuador, and the Cauca Valley and Bogotá area in Colombia, well-doing herds of Holstein-Friesian are seen that have been upgraded from the Criollo. Venezuela has its milkshed of adapted Brown Swiss. In total numbers these dairy herds would probably exceed the number of Criollo now in existence, but in order to avoid confusion their location is not indicated in the diagram.

The Spanish cow played the major role in founding the great herds in the range lands of the Western Hemisphere. This portion of her heritage is solidly established. Another legacy hangs by a thread.

The beautiful and productive Criollo breeds which have been developed by selective breeding in Colombia, Cuba, Costa Rica, and Venezuela are an asset little appreciated by either cattlemen or animal scientists, even in their own countries. The current trend is to transform these breeds, ideally adapted to the tropics, to Zebu-type cattle. Nearly all privately owned herds have already been bred out or are involved in the process. Where the Criollo breeds continue to exist, the cattle for the most part are owned by an experiment station or similar government entity. Some effort is being made, as in Colombia, to maintain Criollo herds to furnish foundation stock for cross-breeding; Cuba is continuing the development of a dual-purpose, milk-beef Criollo in a small segment of her national herd;

Venezuela, with more than 1,000 head of Limoneros, currently has a program to continue the breed. An economy program, the next election, or the whim of an agricultural minister, could eliminate any of the current government breeding programs overnight. The future of the Criollo breeds is far from assured.

The unimproved Criollo, a true survivor of the fittest, is now literally an endangered species. In inhospitable environments under natural selection, these highly specialized types evolved from the same gene pool as the Criollo breeds. The Llanero, the Florida Scrub, the "Spanish" or "Mexican" cattle of Texas before the development of the Longhorn—all made it on their own. For three centuries the only attention they received from man was the roundup at which the best were cut out to be sent to slaughter. These ideally adapted types were small animals, gave barely enough milk for a calf, and produced an inferior carcass, but they lived and multiplied where no other *bos taurus* could exist.

The plant geneticist used some miserable-looking ears of corn to breed his most productive hybrids. When his colleague, involved with the bovine, reaches the same plane of achievement, he will find some building blocks of unknown value have disappeared if the Criollo goes the way of the aurochs. The breeding out of these hardy beasts is a sad page in the story of modern cattle production.

Appendix I. Primary Sources of the Cattle Populations of the Western Hemisphere, 1493 to 1975

The only bovine in the Western Hemisphere before the advent of the European was the North American bison, native to the central United States. This noble beast, a primary food source for the Plains Indians, is now of only minor economic importance, and has been preserved from extinction by the founding of private and government-protected herds. (Another story entirely is the effort, historical and current, to cross the buffalo with domestic cattle.)

The llama, a most useful beast of burden to the Incas, along with its relatives in the Andes, belongs to the camel family. The musk ox of the arctic is a member of the sheep and goat family.

The first cattle on the hemisphere were those brought to the island of Hispaniola by the Spaniards in 1493. In the course of nearly five centuries, a number of other cattle types were introduced and interbreeding occurred as man spread over the habitable areas of the New World. The present cattle population consists of European nonhumped breeds in the temperate zones, derivatives of the Indian Zebu in the tropics, and a few scattered remnants of the Criollo, the pure descendants of the original Spanish cattle.

Cattle entered the Western Hemisphere during six major periods which can be rather narrowly defined. These are:

285

I.   1493–1525   The introduction of Spanish cattle to the Indies
II.   1531–1550   The introduction of Portuguese cattle to Brazil
III. 1608–1650   Northern European pre-breed cattle brought to the eastern seaboard of North America
IV. 1817–1930   Northern European pure breeds enter North America
       1860–1940   Northern European pure breeds enter South America
V.   1835–1854   Zebu cattle appear in United States (nominal)
       1860   Zebu cattle appear in Jamaica
       1870–1930   Zebu cattle appear in Brazil
                     (Zebu breeds frequently reached North and South America from Mexico, having arrived there from Brazil.)
VI. 1966–   The recent introduction of additional European breeds to Canada and United States

These cattle movements have been depicted in the diagram on page 282. The entrance of Spanish cattle to the Indies and their spread throughout the hemisphere has been reviewed in detail. The introduction of Portuguese cattle to Brazil is covered in Appendix II. Following is a summary of the other cattle movements that form the present cattle populations on both continents.

Northern European Pre-Breed Cattle Introduced to North America

The earliest reported movement of cattle to North America from Europe north of the Iberian Peninsula was the landing of a few head on Sable Island off the coast of Nova Scotia in 1518. These were brought from Normandy by the Frenchman, Baron de Levy. Colonization failed, but the cattle existed in a wild state for a number of years before disappearing. There are two other early records of cattle movements from France to Canada. Both were from Brittany, one by Jacques Cartier in 1541 to the province of Quebec, and another to the St. Lawrence area in 1601. Apparently, neither of these introductions survived.

In 1608, Samuel de Champlain brought cattle from Nor-286 mandy to the settlements on the St. Lawrence River. These established the first continuity of a breeding herd in eastern North America. The English established cattle in Virginia a year later, and on the island of Bermuda by 1614. A few cattle may have been landed earlier on this island.

Along with subsequent importations, the Champlain cattle were the foundation stock of the French Canadian (now Canadian) breed which is still maintained as a dairy type in Quebec. The animals that were first landed in Virginia were the forerunners of the Native American cattle. The cattle in Bermuda continued in isolation down to modern times.

The now forgotten "Native" cattle, though unheralded, served the same purpose in building the national herd in the eastern United States that the Longhorn did three centuries later in the West. The Native American comprised the entire cattle population of the United States east of the Missouri River for more than two centuries after the English colonies were founded. These were a conglomerate mixture of the pre-breed types of bovines imported from England, and to a lesser extent, from Holland, Denmark, and Sweden. The Pilgrims brought their cattle from Devonshire, England, in 1620, the same year as the arrival of some Dutch pre-Friesian black-and-white lowland cattle on Manhattan Island. A little later, the Red and Whites were imported. In 1633 the Danes landed their "Yellow" cattle and trailed them to New Hampshire, and in 1638 a few cattle from Sweden were taken to the Swedish settlement in Delaware.

These and subsequent shipments of pre-breed types of cattle inextricably mixed as the pioneers moved westward. They were uncontaminated with any other cattle types until the advent of the purebreds, the forerunners of which arrived early in the nineteenth century.

Some writings on early colonial times mention cattle from the Indies being shipped to ports in Virginia and the Carolinas. During the seventeenth century the Spaniards established missions as far north as Virginia. These undoubtedly had cattle, but it is highly improbable that any of them survived, as all the Spanish outposts were quickly driven out by the British. *287*

Unsubstantiated statements refer to the cowpen people in the Piedmont area of Virginia encountering wild Spanish cattle. Any such infusion of Spanish cattle north of Florida was certainly minor. The Native American and the Spanish cattle were well separated by distance and geographical barriers until the first decades of the twentieth century. The one exception to this generality was in Louisiana where the first French settlements in the early 1700's managed to obtain a nucleus of Spanish cattle from Santo Domingo and Cuba.

Northern European Pure Breeds to the Western Hemisphere

In England and Scotland toward the end of the eighteenth century, intensive methods of selection for type, color, and conformation were practiced on local populations of cattle. This was the beginning of the cattle breeds in the western world as we know them today. Herdbooks were established to record the ancestry of individual animals, and breed societies were organized to set the standards and promote the advantages of one breed over others. The landed gentry of England were the first to systematize the activities of the breed society and the method of handling the herdbook. Husbandmen on the continent soon began to follow the English pattern and applied it to such practices as had grown up among local groups of farmers in selecting breeding stock. These procedures were then formalized in herdbook and breed society.

The introduction of the European breeds to the cattle population of eastern North America began nominally in the early nineteenth century. Fifty years later the European cattle reached the temperate zone of South America. By the end of the century the pure breeds had shaped the destiny of most of the New World cattle outside the tropical and subtropical regions.

The first Shorthorn cattle for which continuity was established were brought to the United States in 1817. A few head had been introduced in the latter part of the eighteenth century, but no record of their progeny was retained. Purebred cattle were the property of only the financially successful farmers for many years, but by 1850 the Shorthorn was well established in a number of localities in the eastern United States and was beginning to move westward. In the late 1850's a few fore-runners of the breed reached southwest Texas.

The Shorthorn reached Canada in 1825, and there were further importations from England over the following two decades. The breed was first used in the East and then in western herds as large-scale ranching operations were organized in the prairie provinces in the last quarter of the nineteenth century.

The Shorthorn and Hereford were the two breeds that were largely used to upgrade the Spanish-type cattle that crossed the Rio Grande. The typical pattern of the early rancher was to use whatever Shorthorn bulls he could lay his hands on. A decade later he switched to Hereford bulls for upgrading the Shorthorn-Spanish crossbreds. There were naturally many variations in the individual rancher's breeding programs, but the Shorthorn gradually disappeared from the range country and the Hereford took over.

Zebu Cattle to the Western Hemisphere

There were several early importations of Zebu cattle to the United States which are of historical interest only. The humped cattle which were to play a substantial role in developing the present cattle population in the tropics of the hemisphere entered several decades later.

The first humped cattle in the Western Hemisphere appear to have been four cows and two bulls imported from Egypt in 1835 by two South Carolina planters. These were African cattle many generations removed from the true Zebu of India. Their descendants soon disappeared. Two Zebu bulls which were imported into Louisiana in 1854 did leave a line of cross-bred progeny which was commonly recognized as better-doing cattle than the local variety. There was no continuity of this breeding, however.

Zebu were introduced to Jamaica in the early 1860's and to Brazil a decade later. Brazil soon became the fountainhead from which tropical America was stocked with the Zebu. The principal breeds which had been received in that country were the Gyr, Guzerat, and Nellore from India. A native type, embodying the genes of these breeds, was later developed and called the Indo-Brazil. All of these breeds were exported from Brazil to hot, tick-plagued areas of the hemisphere. The heaviest shipments went to Mexico and were then transshipped to the United States and northern South America. There were two sizable importations direct to the United States from India, but the quarantine restrictions eventually blocked all shipments except those from Mexico.

Recent Introductions of European Breeds to the United States and Canada

In 1966 Canada opened a major security quarantine station on Grosse Ile in the St. Lawrence River. Soon there began an influx of many European breeds, mostly of beef or dual-purpose type, heretofore largely unknown in Canada and the United States. A few years later the French, under both Canadian and United States auspices, established another quarantine station on the Island of St. Pierre south of Newfoundland, and the imports accelerated throughout the ensuing years.

Much of the activity connected with this movement has been highly promotional, centered on the cattle entrepreneur's exploitation of each new introduction as being superior to anything that preceded it. There is no question, though, that when the European purebred importations were being made in the 1800's, many good breeds were overlooked or were not readily available. Such excellent old cattle types as the Simmental (now making its second appearance on the American scene), the Gelbvieh, and the Limousin, as well as others of the continental breeds, will have a significant impact on the future cattle population of the Western Hemisphere. The story of the Criollo, however, had been concluded before this last infusion of European breeds began.

Appendix II. The Introduction of Cattle to Brazil

Brazil, originally stocked with Portuguese cattle and well isolated from the rest of South America until modern times, does not enter into the story of the Criollo. In order to complete the outline of cattle introductions to the Western Hemisphere, a brief account of the bovine in Brazil is given here.

The papal bull of 1493 first divided the New World between Spain and Portugal by a north-south line drawn 100 leagues west of the Cape Verde Islands. In 1494, by mutual agreement, later sanctified by treaty, the line was moved to 370 leagues west of the islands, roughly to a line between the 48th and 49th meridians. Actually this left three-fourths of Brazil in the area allotted to Spain, but because the Portuguese began their conquest well to the east of the line of demarcation, their advance westward across Brazil proceeded for the most part uncontested.

The Spanish explorer Vicente Yáñez Pinzón had touched land in Brazil early in 1500 without making any claim, probably because he realized he was within Portuguese domain. A few months later Pedro Alvarez Cabral, a Portuguese navigator in command of a large flotilla and far off his course, landed and claimed Brazil for Portugal. Whether, having heard of new lands to the north that were being discovered by the Spaniards, his wandering westward was intentional will probably never be known. He was under orders at the time to follow the newly discovered route around the horn of Africa to the East Indies. Portugal was busy setting up trading posts on her recent discoveries in the Far East, and after a small settlement was established at Salvador in 1502, little attention was given Brazil for another three decades.

Animals of the same type as the progenitors of the Mertolenga breed in present-day Portugal comprised the first cattle which were introduced into Brazil. In 1532, Martim Affonso de Sousa, the recipient of a large land grant from the Crown in the new possession, left Portugal "with 5 ships, 400 crew and colonists, seeds, plants, and domestic animals and started a colony at São Vicente." This was the first permanent settlement in Brazil. Founding an agricultural community was a new approach for Portugal which had been intent only on discovering new lands and establishing trading posts.

There were other expeditions to enlarge this foothold in the new land, and in 1537 another major settlement was established at Olinda, near Recife. São Paulo, in Sousa's captaincy, was the next large town. By the middle of the sixteenth century a prosperous sugar economy was in full swing. Negro slaves were brought from Africa as the rather sparse native population did not take happily to work or subjugation. Colonization was confined to the coastal areas for 150 years, the cattle ranches gradually moving out beyond the cultivated lands. Westward expansion across the country, similar in many respects to the western movement of the pioneers in the United States, began at the end of the seventeenth century with the mining activity in the interior.

Later, reinforcements to the colonies brought cattle from the Cape Verde Islands where Portugal had been established since 1460. This source of imports reduced by one-third the time that animals had to be held aboard ship as compared with loadings in Portugal. The Cape Verde Islands thus served the same function in stocking Brazil as the Canaries had for the Indies.

The early plantation area extended inward from the lengthy coastline between Santos and Recife. Fazendas, large estates on the order of the European medieval feudal manor, expanded their production of sugar, coffee, and tobacco, thus increasing their requirements for draft animals. Cattle ranchers then moved westward from the cultivated areas. The early eighteenth century saw many ranches running from 100 to 1,000 head of cattle. The Mertolenga breed in Portugal is closely related to the Retinto of Spain. In the pre-breed days of the sixteenth century, when Spain and Portugal were stocking their new possessions, there was probably little difference in the animals that each country took to its overseas colonies.

For three and a half centuries after their introduction, the Portuguese cattle spread over the inhabited areas of Brazil without any significant mixing with other types. British Guiana was stocked with cattle from Brazil which came through the Rupununi district from the Amazon basin. This is the only other region in South America where the cattle population stemmed from Portuguese rather than Spanish stock. Contact with Spanish cattle along the borders of Bolivia, Paraguay, and Argentina were only of minor extent.

When the Zebu were introduced to Brazil in 1870, crossing with the Portuguese cattle proceeded rapidly. Little, if any, effort was made to perpetuate the old type of cattle. Except for the dairy herds of European breeds around the large cities and the water buffalo in the northeast corner, the bovine population is practically all Zebu or Zebu-related types such as the Santa Gertrudis.

SOURCES CONSULTED

Allen, Lewis F. *American Cattle: Their History, Breeding and Management.* New York, O. Judd Company, 1883.

Archivo General de Indias, Sevilla, Spain. Archives, from time of Columbus to 1530.

Arnade, Charles W. "Cattle Raising in Spanish Florida—1513–1763" *Agricultural History*, Vol. XXXV, No. 3. Published for the Agricultural History Society at the Garrard Press, Champaign, Illinois, 1961.

Bannon, John Francis. *The Spanish Borderlands Frontier.* New York, Holt, Rinehart and Winston, 1970.

Bidwell, Percy Wells, and John I. Falconer. *History of Agriculture in the Northern United States, 1620–1860.* Washington, Carnegie Institution, 1925.

Bishko, Charles Julian. "The Peninsular Background of Latin American Cattle Ranching," *The Hispanic American Historical Review*, Vol. XXXII, No. 4, November, 1952.

Boussingault, J. B., y François Désiré Roulin. *Viajes Científicos a los Andes Ecuatoriales.* Paris, Librería Castellana, 1849.

Brand, Donald D. "The Early History of the Range Cattle Industry in Northern Mexico," *Agricultural History*, Vol. XXXV, No. 3. Published for the Agricultural History Society at the Garrard Press, Champaign, Illinois, 1961.

Brinton, Daniel G. *Notes of the Floridian Peninsula.* Philadelphia, Joseph Sabin, 1859.

Burcham, L. T. "Cattle and Range Forage in California: 1770–1880," *Agricultural History*, Vol. XXXV, No. 3. Published for the Agricultural History Society at the Garrard Press, Champaign, Illinois, 1961.

Burns, E. Bradford. *A History of Brazil.* New York, Columbia University Press, 1970.

Camp, Paul D. A Study of Range Cattle Management in Alachua County, Florida. Bulletin 248. Gainsville, Agricultural Experiment Station, 1932.

Collecion de documentos inéditos relativos al descubrimiento, conquista y colonizacion de las posesiones españolas en América y Oceanía. Edited by Luis Torres de Mendoza, y otros. Madrid, 1864.

Conner, Seymour V. *Texas, a History.* New York, Thomas Y. Crowell Company, 1971.

Cox, James. *Historical and Biographical Record of the Cattle Industry and the Cattlemen of Texas and Adjacent Territory.* St. Louis, Woodward & Tiernan Printing Company, 1895.

Cross, Joe. *Cattle Clatter.* Kansas City, Walker Publications, 1938.

De Alba, Jorge. "Observations on the Criollo Breeds." Unpublished monograph. Library of Congress, Washington. 1955.

———, and Candelario Carrera. *Communicaciones de Turrialba,* No. 61. Turrialba, 1958.

Denhardt, Robert M. *The Horse of the Americas.* Norman, University of Oklahoma Press, 1975.

Díaz del Castillo, Bernal. *The Discovery and Conquest of Mexico.* New York, Farrar, Straus and Cudahy, 1956.

Dobie, J. Frank. *The Longhorns.* Boston, Little, Brown and Company, 1941.

Dominguez, Luis L., ed. *The Conquest of the River Plate 1535–1555.* New York, Burt Franklin, n.d.

Duke, William. *Memoirs of the First Settlement of the Island of Barbados.* Boston, Wm. Beeby, 1741.

Dunbar, Gary S. "Colonial Carolina Cowpens," *Agricultural History,* Vol. XXXV, No. 3. Published for the Agricultural History Society at the Garrard Press, Champaign, Illinois, 1961.

Dusenberry, William H. *The Mexican Mesta; the Administration of Ranching in Colonial Mexico.* Urbana, University of Illinois Press, 1963.

Fairbanks, George R. *History of Florida.* Philadelphia, J. B. Lippincott, 1871.

Ferris, Robert G., ed. *Explorers and Settlers.* Washington, Government Printing Office, 1968.

Frere, George. *A Short History of Barbados.* London, J. Dodsley, 1768.

Fugate, Francis L. "Origins of the Range Cattle Era in South Texas," *Agricultural History,* Vol. XXXV, No. 3. Published for the Agricultural History Society at the Garrard Press, Champaign, Illinois, 1961.

García Mercadal, José, *Lo que España llevó a America.* Madrid, Taurus, 1959.

Godlet, Theodore L. *Bermuda.* London, Smith Elder and Company, 1860.

Gray, Lewis Cecil. *History of Agriculture in the Southern United States to 1860.* Washington, Carnegie Institution, 1933.

Hamilton, Earl J. *American Treasure and the Price Revolution in Spain, 1501–1650.* New York, Octagon Books, 1965.

Haring, Clarence Henry. *The Spanish Empire in America.* New York, Oxford University Press, 1947.

Hemming, John. *The Conquest of the Incas.* New York, Harcourt Brace, Jovanovich, 1970.

Herring, Hubert Clinton. *A History of Latin America.* New York, Alfred A. Knopf, 1955.

Humboldt, Alexander von. *Personal Narrative of Travels to the Equinoctial Regions of the New Continent.* London, printed for Longman, Hurst, Reis, Orme & Brown, 1814.

Huntington, Henry E., Library & Art Gallery Catalog. *From Panama to Peru.* London, Maggs Brothers, 1925.

Irving, Theodore. *The Conquest of Florida.* London, Edward Churton, 1835.

Isaza, Jaime R. "Monografia Sobre Razas Criollos en Colombia." Unpublished monograph. Library of Congress, Washington. January, 1956.

James, Preston Everett. *Latin America.* 4th ed. New York, Odyssey Press, 1969.

Johnson, Willis Fletcher. *The History of Cuba.* New York, B. F. Buck and Company, 1920.

Kidd, K. K. *Biochemical Polymorphismus, Breed Relationships, and Germ Plasma Resources in Domestic Cattle.* Madrid, Proceedings Conference on Animal Reproduction, 1974.

Lefroy, J. Henry. *The History of the Bermudas.* New York, Burt Franklin, 1882.

Lydekker, Richard. *The Ox and Its Kindred.* London, Methuen and Company, 1911.

Madariaga y Rojo, Salvador de. *Christopher Columbus.* New York, Frederick Ungar Publishing Company, 1942.

Major, Richard Henry. *Select Letters of Christopher Columbus.* London, printed for the Hakluyt Society, 1847.

Martínez-Hidalgo, María José. *Columbus' Ships.* Edited by Howard I. Chapelle. Barre, Mass., Barre Publishers, 1966.

Mason, Ian Lauder. *A World Dictionary of Livestock Breeds, Types and Varieties.* Farnham Royal, England, Commonwealth Agricultural Bureau, 1969.

Mesa, Daniel Bernal. "Historia de las Primeras Importaciónes de Ganado a Colombia." Unpublished monograph. Library of Congress, c.1965.

Morison, Samuel Eliot. *A Life of Christopher Columbus, Admiral of the Ocean Sea.* Boston, Little, Brown and Company, 1942.

Morrisey, R. J., "Northward Expansion of Cattle Ranching in New Spain, 1550–1600" *Agricultural History*, Vol. XXXV. Published for the Agricultural History Society at the Garrard Press, Champaign, Illinois, 1961.

Patiño, Victor Manuel. *Plantas Cultivades y Animales Domésticos en América Equinoccial.* Vol. V. Cali, Impreta Departamental, 1970.

Pinzón, Emigdio M. *Ganados Nativos Colombianos.* Bogotá, Industria Colombiana, Impreta Departmental, 1955.

———, and others. "Bovinos Criollos Colombianos," *Boletín de Divulgación*, No. 5. Bogotá, Ministerio de Agricultura de Colombia, 1959.

Post, Lauren C. *The Old Cattle Industry of Southwest Louisiana.* Reprinted from *McNeese Review*, Vol. IX, 1957.

Prescott, William Hickling. *History of the Conquest of Mexico.* New York, A. L. Burt Company, 1843.

———. *History of the Conquest of Peru.* Philadelphia, J. B. Lippincott, 1847.

———. *History of the Reign of Ferdinand and Isabella.* 3rd ed. New York, A. L. Burt Company, 1838.

Puente y Olea, Manuel de la. *Los Trabajos Geográphicas de la Casa de Contratación.* Sevilla, 1910.

Read, Jan. *The Moors in Spain and Portugal.* London, Faber & Faber, 1974.

Rouse, John E. *World Cattle.* Norman, University of Oklahoma Press, 1970.

———. *Cattle of North America.* Norman, University of Oklahoma

Press, 1972.

Sauer, Carl Ortwin. *The Early Spanish Main.* Berkeley and Los Angeles, University of California Press, 1966.

Smith, John. *The Generall Historie of Virginia, New-England, and the Summer Isles.* Printed by I. D. and I. H. for Michael Sparkes, n.p., 1624.

Thompson, James Westfall. *History of Livestock Raising in the United States, 1607–1860.* Report of U.S. Department of Agriculture. Washington, Government Printing Office, 1942.

United States Bureau of Animal Industry. *Eighteenth Annual Report, 1901.* Washington, Government Printing Office, 1902.

Vázquez, Ignacio Munoz. "Aportacion a la Historia de la Veterinaria." Unpublished monograph. Library of Congress, Washington. 1974.

Velásquez, José Q., *Ganada Blanco Orejinegro* Bogotá, Publicacion del Banco Cafetero, n.d.

Warren, Harris Gaylord. *Paraguay, an Informal History.* Norman, University of Oklahoma Press, 1949.

White, Charles Langdon. "Cattle Raising a Way of Life in the Venezuelan Llanos," *Scientific Monthly*, September, 1956.

Wilkinson, Henry. *The Adventurers of Bermuda.* London, Oxford University Press, 1933.

Winsor, Justin. *Spanish Exploration and Settlement in America from the 15th to the 19th Century.* New York, Houghton Mifflin Company, 1886.

Youatt, William. *Cattle.* New York, C. M. Saxton, 1851.

INDEX

There's nothing very beautiful and nothing very gay
About the rush of faces in the town by day,
But a light tan cow in a pale green mead,
That is very beautiful, beautiful indeed.

Orrick Johns

Surinam

French Guiana

Antigua

El Salvador

Cuba

Florida

Oklahoma